DELUGE

Ireland's weather disasters
2009–2010

Kieran Hickey

OPEN AIR

Typeset in 11pt on 13pt Ehrhardt
by Carrigboy Typesetting Services for
OPEN AIR an imprint of FOUR COURTS PRESS LTD
7 Malpas Street, Dublin 8, Ireland
www.fourcourtspress.ie
and in North America for
FOUR COURTS PRESS
c/o ISBS, 920 N.E. 58th Avenue, Suite 300, Portland, OR 97213.

A catalogue record for this title is available
from the British Library.

ISBN 978–1–84682–271–1

This publication was grant-aided by the Publications Fund of
National University of Ireland, Galway

NUI Galway
OÉ Gaillimh

Printed in England
by MPG Books, Bodmin, Cornwall.

For Deirdre, Michael and Sean

Contents

Figures, tables and plates

COLOUR PLATES
(between page 64 and page 65)

Acknowledgments

This book would not have been published without a grant-in-aid provided by the Publications Fund of the National University of Ireland, Galway. This fund plays a very important role in supporting publications in a wide range of areas. My thanks also go to my colleagues in the Department of Geography, NUI Galway, and elsewhere in the university and other universities in Ireland and abroad for their support throughout the years. A special thanks to Four Courts Press for accepting this proposal and especially to the editorial team of Michael Potterton and Martin Fanning, who had the unenviable task of converting my writing into a viable publication, and all within a very tight schedule. My thanks also to Met Éireann for permission to use a number of key diagrams throughout the book, and to a number of organisations and individuals for allowing publication of various photographs, including Keith Lambkin of Valentia Observatory, the Central Fisheries Board, the *Irish Examiner*, NASA Earth Observatory, the Office of Public Works, Rob Fisher, Gary Fox, Sigurlaug Gunnlaugsdóttir, Anne Kearney, Claire Lyons, Tony Murphy, William Murphy and Ólafur Sigurjónsson. Every effort was made to identify and contact copyright-holders for permission. All errors and omissions are my responsibility, but hopefully these are few.

Galway, August 2010

Abbreviations

AA	Automobile Association
ERC	Emergency Response Committee
ESB	Electricity Supply Board
EU	European Union
FBD	Farmer Business Developments
FEMA	Federal Emergency Management Agency
GBCO	Great Britain Census Office
HSE	Health Service Executive
IBEC	Irish Business and Employers Confederation
IFA	Irish Farmers' Association
IIF	Irish Insurance Federation
MEP	Member of the European Parliament
NAMA	National Asset Management Agency
NASA	National Aeronautics and Space Administration
NCB	National City Brokers
NIE	Northern Ireland Electricity
NUI	National University of Ireland
OPW	Office of Public Works
PSNI	Police Service of Northern Ireland
TD	Teachta Dála (member of Dáil Éireann)
UCC	University College Cork
WMO	World Meteorological Organization

CHAPTER I

Introduction

The twelve months from the summer of 2009 to the summer of 2010 were exceptional in terms of weather. During this period, Ireland experienced weather disasters not seen for a generation or more, occurring one after another. In addition, three rare geophysical events affected the country. This book explains what happened, the rarity of the events within the context of previous occurrences of similar nature, and their dramatic impacts, which cannot be underestimated. The role of humans in exacerbating the severity of the events is also considered and the book demonstrates just how oblivious we have become to the natural world and its vagaries.

The summer of 2007 was unusually wet. It was very disappointing, but the expectation was that it would be a one-off in the greater scheme of things. This expectation was proven wrong, however, as the summer of 2008 was wetter and the summer of 2009 was worse again, with new summer rainfall records being set. There was some relief from the almost constant rainfall of the summer of 2009 in September and October of that year, but this was short-lived. Along came November, which turned out to be not only the wettest November on record for many meteorological stations around Ireland but also the wettest month *ever* recorded for some stations, several of which had been in operation since the 1800s.

The exceptional rainfall of 18–20 November was the trigger for some of the worst flooding to have affected Ireland for thirty years, and for much longer in the west of Ireland. Flooding affected a huge part of the country, but the worst affected county was Galway, where the flooding was still evident in places in February 2010. Vast tracts of countryside and numerous villages and towns throughout the county were flooded. Many of these areas included farmhouses as well as new and relatively new housing estates. The damage and desolation of the flood victims was shocking in the extreme. Remarkably, Galway city was relatively unaffected by floodwaters, being well-drained by the Corrib. But the city was effectively cut-off from the outside world as all the major route ways out of Galway were impassable for several days. This included the railway as well. Only for the temporary and emergency opening of stretches of the new Galway to Dublin motorway, this situation would have lasted much longer.

Cork city suffered one of its biggest flood events ever, but the cause here was indirectly related to the spectacular rainfall of 18–20 November. Conditions at the Electricity Supply Board (ESB) dam at Inniscarra were such that the amount of water being released had to be continually increased, leading to the flooding of the city. This occurred because of rising water levels in the reservoir behind the dam, and the fear of the dam being over-topped and the possibility of dam failure. In the latter case, this would have been catastrophic and could have led to fatalities, particularly as nearly everybody was asleep in bed when the flooding occurred. Conditions in parts of Co. Cork were not much better.

Massive flooding also occurred on the Shannon and again the operation of the Parteen Dam played a crucial role in this. The ESB was faced with a similar scenario to that at Cork and ultimately had to release excess water down the system, leading to flooding. Many of the towns downstream suffered extensive flooding and once again some new and nearly new housing estates were badly affected, as well as huge tracts of agricultural land. Flooding was not confined to these locations and serious flooding occurred in Cos. Wicklow, Wexford, Waterford, Carlow, Kilkenny, Kerry, Limerick, Clare and Tipperary, where Clonmel once again got a battering. There was also spot-flooding throughout much of the rest of Ireland, but not quite of the severity of the southern half of the country.

There was a remarkable response to the flooding at local authority level and by the many voluntary agencies and eventually by central government. A number of issues emerged and of particular concern was the extent of flooding on many housing estates and how building could have taken place in these locations if they were that vulnerable to flooding. In addition, the issue of flood-control emerged, involving the ESB and numerous other agencies which in some way have an involvement in the management of Ireland's waterways. No clear control-system was evident as a result of being tested by this flood event and there were other political consider-ations as well.

One of the remarkable aspects of the flooding in Co. Galway was its longevity, even on key main roads. There were still diversions in February on the main Galway–Limerick road at Labane on the Galway side of Gort, and this was also the case in some parts of rural south and east Galway and even some of the minor roads. Gort was affected once again by very serious flooding, as not enough had been done to prevent this since previous events (most notably in 1995).

By February the country was starting to come out of the worst cold spell to have affected Ireland since the early 1960s, which started in late

December and carried on until mid-January. It could be argued, however, that the cold spell, although of much less severity and not continuous, stayed with us up until early May. It is a very long time since we had such a run of colder than average months.

The effects of the cold spell, with frequent snowfall but little snow-lying, were as dramatic as the flooding. Not least is the fact that it affected the whole island of Ireland, north and south. The country ground to a halt as temperatures dipped below freezing and then below −10°C and there were sustained sub-zero temperatures of this nature for several weeks. Many thousands of people had terrible falls on the frozen ground, resulting in broken bones and other serious injuries. The roads and pathways froze and transport was very difficult. The country virtually ran out of gritting materials and energy consumption reached a record peak as people tried to keep warm.

Spare a thought for some parts of Co. Galway, where the flood waters had yet to fully recede and now were frozen, including at Labane and Kiltartan on the main Galway–Limerick road. Many people were still in the process of trying to fix their houses when the cold spell came along. Of course, when the main thaw came in mid-January, new flooding occurred as a result of severe rainfall and melting ice and snow and thousands of homes and other buildings suffered flooding as a result of burst pipes caused by the cold spell.

The role of central government and the Emergency Response Committee (ERC) was again the focus of much debate and some anger and the issue of gritting supplies became a key discussion point. There was also the absurdity that if you cleared the path in front of your house or property you could be liable if someone were to fall and injure themselves, even though there was much need for path clearance for the common good. Legislation is needed to resolve this.

One would have been forgiven for thinking at this stage that nature had thrown everything it had got at us and that surely there would be some let-up. Maybe the meteorite fall somewhere in the northern half of the country in early February should have been a warning that things were going to get even stranger.

As a result of our own weather crisis, nobody paid much attention to a small volcanic eruption in Iceland in March. The eruption was short-lived and barely made the papers. However, this was little more than the volcano clearing its throat, because on 14 April a large new eruption took place under the icecap of Eyjafjallajökull, pumping vast amounts of steam and, more importantly, volcanic ash into the atmosphere. This was blown over

Europe and caused an initial week-long shut-down of air transport, causing immense travel disruption and cost to the airlines. Further shut-downs occurred when the ash was blown over Europe up until 24 May, when this eruption ceased.

Nature had one more trick to show us, and although it was not meteorological, it is worthy of inclusion in a year which seems to have had everything. This was the Co. Clare earthquake, which occurred on 6 May and was measured at 2.7 on the Richter Scale. The earthquake was noticed by a lot of people and frightened many. It was the first instrumentally recorded earthquake in Co. Clare since records began in 1978.

This book outlines the timeline of this remarkable twelve-month period in the weather history of Ireland and the key events that occurred. It provides a context so that people can understand both the severity of the events and how rare they are and the absolute rarity of such a combination of weather and geophysical disasters over such a short period.

CHAPTER 2

A wet summer again

INTRODUCTION

One of the features of Irish summers in 2007–9 was the persistent rainfall that occurred, giving three of the wettest summers since instrumental records began. Each summer had a different pattern in terms of the rainfall, but the net effects were much the same; persistent rainfall with near record totals. This chapter explores the last three summers and puts them into the context of the flooding that occurred in November 2009 and the role that the summer wetness played in this event. To put the severity of the wetness into context, it is worth noting that June 2010 was the first summer month with below average rainfall in Ireland since August 2006.

THE SUMMER OF 2007

The summer of 2007 was in total contrast to the great summer of 2006. It was the wettest summer for around nine years across much of the country but in the east it was the wettest for fifty years, being wettest in Leinster. Two stations had their wettest summer since records there began: Kilkenny (since 1954) and Casement Aerodrome (since 1957). Rainfall occurred on two out of every three days on average and sometimes for many days in a row. Dublin recorded 33 rain-days (a rain-day is one with 0.2mm of rainfall or more) between 11 June and 29 July (Met Éireann, 2007).

In June, Johnstown Castle recorded 167.2mm, which was the highest of the listed stations and represented 274 per cent of the 1961 to 1990 average rainfall for that month. Nearly all stations recorded over 100mm of rainfall, except Malin Head, and all recorded well above average rainfall, varying from 128 per cent at Malin Head to 274 per cent at Johnstown Castle. There was more variability in the rainfall for July, with NUI Galway recording the highest at 162.3mm, but Casement Aerodrome recorded over three times its normal July rainfall, at 309 per cent (Table 2.1). Only two stations (Valentia Observatory and Malin Head) recorded less than 150 per cent of normal rainfall. July was the worst of the summer months for rainfall, with an improvement in August. Johnstown Castle recorded the highest monthly rainfall total (112.6mm), but Casement Aerodrome again had the highest figure when compared to the long-term average (156 per cent).

The summer of 2008 maintained the previous summer's excessive rainfall figures. Rainfall was well above normal all across the country, with more than twice the average in the east and southeast. Two exceptional results stand out: it was the wettest summer since the foundation of Cork Airport in 1962 and it was the wettest summer at Dublin Airport since 1958. Nearly half of all summer days in most locations had rainfall in excess of 0.2mm (i.e. a rain-day), with the exception of parts of the west and northwest (Met Éireann, 2008).

Table 2.1 Comparison of rainfall totals and % of average (1961–90) for the summer of 2007 (after Met Éireann, 2007).

Station	June		July		August	
	Total (mm)	% of average	Total (mm)	% of average	Total (mm)	% of average
Belmullet	71.9	107	102	150	88.9	95
Casement Aerodrome	137.5	259	145.4	309	107.5	156
Cork Airport	156.9	231	116.3	179	80.4	89
Dublin Airport	134.4	239	119.1	238	95.5	135
Johnstown Castle	167.2	274	85.8	151	112.6	136
Malin Head	83.3	128	105.4	146	86.2	94
Mullingar	108.2	161	150.1	246	108.2	130
NUI Galway	128.3	171	162.3	239	101	100
Shannon Airport	107	170	93.5	164	100.8	123
Valentia Observatory	153.6	192	76.3	105	106	95

The maximum rainfall in June was recorded at NUI Galway (194.7mm). This represented 260 per cent of normal and was also the highest percentage value (Table 2.2). Conditions deteriorated in July, with the maximum rainfall of 160.6mm being recorded at Johnstown Castle, which also had the highest percentage value of 282 per cent, nearly three times normal summer rainfall. July also had one exceptional twenty-four-hour fall, when 90mm was recorded at Castlemahon, Co. Limerick, on the 31st. Conditions were much the same in August, as the summer's excessive precipitation continued. NUI Galway recorded 240.2mm, but was beaten to the highest percentage, which was 271 at Dublin Airport. Five of the ten stations listed had more than twice their normal summer rainfall. There were two exceptional twenty-four-hour falls during this month. On the 9th, Dublin Airport had 76mm and on the 13th, Belderrig in Co. Mayo reached three figures, with 100mm of rainfall (Met Éireann, 2009). We all hoped that surely the next summer would not be wet again; alas we were to be disappointed, as the next section shows.

Table 2.2 Comparison of rainfall totals and % of average (1961–90) for the summer of 2008 (after Met Éireann, 2008; Lennon and Walsh, 2008).

	June		July		August	
Station	*Total (mm)*	*% of average*	*Total (mm)*	*% of average*	*Total (mm)*	*% of average*
Belmullet	127.6	190	44.3	65	192.1	204
Casement Aerodrome	70.5	133	101.7	216	175	254
Cork Airport	138.5	204	148.1	228	163.6	182
Dublin Airport	76.4	136	111.5	223	192.4	**271**
Johnstown Castle	137	225	**160.6**	282	144.8	174
Malin Head	74.8	115	125.8	175	123.2	134
Mullingar	88.8	134	109.3	178	159.4	196
NUI Galway	**194.7**	**260**	84.2	124	**240.2**	238
Shannon Airport	110.8	176	100.8	177	176.6	215
Valentia Observatory	147.8	185	92.8	127	181.7	164

THE SUMMER OF 2009

Rainfall in the northern third of the country was slightly above normal for this summer, but this was the exception and not the rule. For the rest of the country, summer rainfall was from 125 per cent of normal right up to 250 per cent of normal (Met Éireann, 2009). In the latter case, this was a small part of the midlands and the southwest and southeast corners of Ireland. For some Leinster stations, this was the third summer in a row where twice the normal amount was received. One of the features of this wet summer was the almost daily occurrence of rainfall, particularly in July and August. Typically, between sixty and seventy-five rain-days were recorded. In addition, there were between 40 and 60 wet days (a wet day has a higher threshold of 1mm or more of rainfall). These totals were the highest recorded for the southern half of the country since 1985 (Met Éireann, 2009).

In the case of Valentia Observatory at the southwest corner of Ireland, a new summer record was set at 620mm, more than any previous total going back to 1892. This was the first time the 500mm threshold was exceeded, not to mention the 600mm threshold, and was over 150mm higher than the previous record total, which occurred in 1950 (Figure 2.1). The station at Mullingar also recorded its wettest summer since it was founded in 1950 (Table 2.3). Other stations in the southeast recorded summer rainfall totals that had not been reached for fifty years or more, including Foulkesmills in Co. Wexford, where 423mm were recorded for the summer (Met Éireann, 2009).

Table 2.3 Comparison of rainfall totals and % of average (1961–90) for the summer of 2009 (after Met Éireann, 2009).

Station	June		July		August	
	Total (mm)	*% of average*	*Total (mm)*	*% of average*	*Total (mm)*	*% of average*
Belmullet	49	73	113.3	167	180.8	192
Casement Aerodrome	78.2	**148**	105.7	225	67.8	98
Cork Airport	84.8	125	203.7	313	155.5	173
Dublin Airport	76.2	136	153.9	308	69.1	97
Johnstown Castle	84.5	139	215.3	378	167.8	196
Malin Head	49.3	76	70.6	98	165.9	180
Mullingar	90.6	137	192.4	311	135.1	162
NUI Galway	49.2	66	161.1	237	215.7	214
Shannon Airport	58.5	93	116.1	204	119.3	145
Valentia Observatory	**98**	123	**252.1**	345	**274**	**247**

Remarkably, the summer had slightly higher than average temperatures when compared to the 1961–90 average, but it was the coldest summer since 2002. Temperatures in early June were particularly high for that time of the year, with 28.6°C being recorded at NUI Galway on 2 June. This was great for Galway, as it coincided with the Volvo Ocean Race and for the two weeks that the race was in port there was hardly a cloud to be seen. Everyone hoped that this would be the pattern for the summer. Even more bizarre was the fact that sunshine totals were also above average, mostly as a result of the exceptionally sunny June. This was the sunniest June for at least fourteen years and for more than fifty years in the west and north. It was the sunniest summer in the extreme west-coast stations of Malin Head and Belmullet since 1968. Alas, the relatively high temperatures and glorious sunshine disappeared in July and August when days with sunshine were very rare indeed (Met Éireann, 2009).

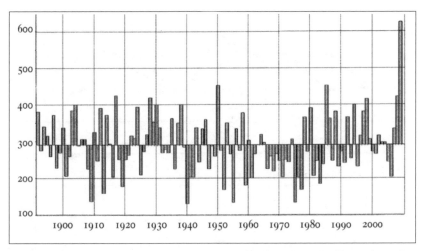

Figure 2.1 Summer rainfall totals (in mm) at Valentia Observatory, Co. Kerry, from 1892 to 2009. The mean value for this period is 290mm (after Met Éireann, 2009).

COMPARISON OF THE SUMMERS OF 2006–9

When the rainfall of the summers of the four years from 2006 to 2009 is compared, it shows the severity of the rainfall levels of the latter three years. The summer of 2006 was the driest, warmest and sunniest since 1995. This is most obvious for rainfall, with all ten stations recording values well below the normal summer rainfall, varying from 44 per cent of normal at Shannon Airport to 88 per cent of normal at Malin Head (Figure 2.2). This was the driest summer for Shannon Airport since records began in 1945. What a contrast with the three following years, when rainfall was well above the normal summer percentage values, varying from 115 per cent (recorded at both Malin Head and Belmullet) to 238 per cent (recorded at both Johnstown Castle and Valentia Observatory) in 2009. However, most stations recorded values between 150 per cent and 225 per cent of normal during these three summers; well above normal and approaching two or two-and-a-half times the usual rainfall.

CONCLUSIONS

One of the major points to emerge with regard to the persistent exceptionally high summer rainfall is that it does not fit into the global warming pre-

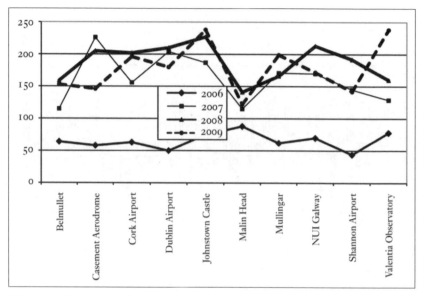

Figure 2.2 Summer rainfall, 2006–9, compared with average summer rainfall, 1961–90 (after Met Éireann, 2006–9).

dictions for Ireland. The predictions suggest a significantly reduced summer rainfall, with summers getting drier as the century progresses. Whether the wet summers of 2007, 2008 and 2009 are just an unusual coincidence or something more systematic and unexpected has yet to be seen. The next few summers will tell (Hickey, 2008). June 2010 was dry and sunny, but all hopes of a glorious summer were dashed by a consistently wet July (Met Éireann, 2010). One of the implications of the 2009 summer in particular is that the normal decrease in river and lake levels and the reduction in the height of the water table did not occur. This means that when rainfall levels started to pick up in the autumn there was little spare capacity in the rivers and lakes and in the ground. This is an ideal scenario for the generation of a major flood. The only surprise is that it did not happen in the previous two autumns, as conditions then were much the same. The timing and amount of precipitation in autumn was a crucial factor in whether there was a major flood or not.

CHAPTER 3

The November 2009 deluge and flooding

INTRODUCTION

This chapter looks at the lead-up to the flooding which started in mid-November and shows that the high water levels in river channels and lakes were maintained as well as a very high groundwater level. The straw that broke the camel's back was the exceptional rainfall that occurred over the forty-eight-hour period between 18 and 20 November (pl. 1). The flooding that followed was on a scale not previously seen or expected by anyone. December rainfall was sufficient to sustain the flooding in the west and in Co. Galway in particular.

SEPTEMBER AND OCTOBER RAINFALL

The constant rainfall of the summer continued into September and no real change occurred from the 9th onwards. Most of the rainfall recorded during the month almost everywhere occurred in the first eight days. There was even some localised flooding in some parts of the country as a result. This was a bad sign leading into the winter and the wetter half of

Table 3.1 Comparison of rainfall totals and % of average (1961–90) for September and October 2009 (after Met Éireann, 2009).

Station	September		October	
	Total (mm)	*% of average*	*Total (mm)*	*% of average*
Belmullet	45.6	42	114.5	85
Casement Aerodrome	28.8	46	103	149
Cork Airport	53.8	55	180.6	143
Dublin Airport	24.2	n/a	84.9	n/a
Johnstown Castle	51.6	n/a	179.2	166
Malin Head	71	70	118.5	100
Mullingar	38.1	n/a	111.7	119
NUI Galway	63.9	59	113.4	88
Shannon Airport	56.9	69	106.9	115
Valentia Observatory	90	72	230.9	147

the year, and was a warning that there was little spare capacity in the ground and the water bodies around the country. The month then went to the other extreme, with well below average rainfall and most stations recording their driest September since 2002. At Claremorris, it was the driest September since 1990. None of the selected stations exceeded 90mm of rain or 72 per cent of normal (Table 3.1). Remarkably, even the sun returned. Critically, however, there was enough rainfall over the country to maintain high water levels in rivers and lakes and also to maintain the high water table.

October saw a return to heavier rainfall, which occurred in patches throughout the month but particularly on the 6th and 9th, from the 19th to the 24th and on the 30th and 31st. On the latter dates, forty-eight-hour rainfall totals in excess of 50mm were recorded in parts of the east and south and this resulted in some localised flooding, including in Inistioge in Co. Kilkenny. The actual totals were lower than normal in the west and north and above normal in the south and east. Valentia Observatory recorded 230mm of rainfall and Cork Airport had 180.6mm, but the rest of the stations were between 100 and 119mm (Table 3.1). Johnstown Castle in Co. Wexford had 179.2mm of rainfall, its highest October total since 1982. This was the highest value, at 166 per cent, but the rest of the values were either close to normal or below 150 per cent of normal (Met Éireann, 2009). There was again more than sufficient rainfall, particularly in the south, to maintain high water levels in the ground, rivers and lakes. The localised flooding so early in the year clearly shows the lack of capacity that was already a problem.

NOVEMBER RAINFALL

The rainfall in November that led directly to the flooding was exceptional in many respects and many records were broken as a result (pl. 2). The country was affected by a whole series of depressions which crossed the country repeatedly throughout the month, bringing with them regular and sustained high falls of rain. Particularly heavy falls occurred on the 1st, 9th, 16th to 19th and 21st in Connacht and Munster. In addition, a heavy fall on the 29th brought flooding to the east as well as the main flood, which was triggered by the downpour of the 16th to 19th, as we will see (Met Éireann, 2009).

The rainfall totals were astronomical and were the highest on record at most stations throughout the country. Valentia Observatory had a total of 345.4mm (235 per cent of normal), and this was the highest of any month since observations began in 1866 (Table 3.2). Similarly, the meteorological station at NUI Galway recorded 329.4mm (286 per cent of normal), and this was the highest of any month since 1881. Nearly one-third of a normal year's rain fell in one month in Galway. There seemed to be no let-up, and rain seemed to fall for days at a time. It also seemed incredibly dark as heavy rain clouds appeared to stay overhead persistently.

Table 3.2 Comparison of rainfall totals and % of average (1961–90) for November 2009 (after Met Éireann, 2009).

Station	November	
	Total (mm)	*% of average*
Belmullet	208.5	164
Casement Aerodrome	169.0	256
Cork Airport	232.8	214
Dublin Airport	149.4	n/a
Johnstown Castle	205.8	187
Malin Head	217.7	189
Mullingar	191.6	204
NUI Galway	329.4	**286**
Shannon Airport	245.6	259
Valentia Observatory	**345.4**	235

A number of rainfall stations at high altitude in Co. Kerry recorded values in excess of 600mm, including 649mm at Cloone Lake. Three climato-logical stations at altitude recorded remarkable rainfall totals in excess of 400mm (Delphi Lodge, Co. Mayo, with 465mm; Carron in the Burren, Co. Clare, with 438.9mm; and Maam Valley in Connemara, Co. Galway, with 452.6mm). None of these stations has been in existence long enough to calculate the percentage of normal values for that month. Sherkin Island had the highest value, at 292 per cent or nearly three times the normal monthly rainfall for November (Met Éireann, 2009). The stations recorded values of between 164 per cent and 286 per cent of normal, with most stations having over twice their normal rainfall for November (Table 3.2).

One way of examining the extremes of rainfall that occurred is to look at their recurrence interval. This measure identifies how often a particular value of rainfall is likely to occur. The bigger the value of the rainfall, the longer the return period. So, for parts of Cos. Galway, Roscommon, Sligo, Clare and Leitrim, the recurrence interval was between 150 and in excess of 500 years, indicating that these rainfall amounts should only recur between once every hundred-and-fifty and once every five hundred years (Table 3.3). This indicates the severity of the rainfall (McGreevy, 2010a).

Other areas had values into the hundreds of years. Of course, care must be taken when using recurrence intervals; they are not predictive, as two fifty-year values could occur in a single month and then not for the next two hundred years, as the recurrence interval is only a statistical averaging tool (Walsh, 2010).

Table 3.3 Exceptional Recurrence Intervals (above 300 years) of 25-day rainfall totals in November 2009 (after Walsh, 2010).

Station	County	Total (mm)	End Date	Return period (years)
Drumconnick	Cavan	231	25th	346
Kilmaley	Clare	460	26th	>500
Doo-Lough	Clare	352	25th	>500
Killaloe	Clare	371	26th	>500
Tulla	Clare	342	26th	>500
Ballinasloe	Galway	271	26th	>500
Gort	Galway	413	26th	>500
North Kerry Landfill	Kerry	365	26th	341
Drumshanbo	Leitrim	318	25th	>500
Aughnasheelan	Leitrim	369	25th	471
Drumsna	Roscommon	278	26th	>500
Boyle	Roscommon	287	26th	>500
Elphin	Roscommon	279	26th	>500
Frenchpark	Roscommon	320	25th	>500
Ardtarmon	Sligo	250	25th	>500

The rainfall of 18–20 November
The major flooding eventually came after the exceptional rainfalls that occurred between 16 and 20 November, with over 100mm of rain falling in forty-eight hours in many south-western areas between the 18th and 20th (pl. 1). In fact, some of the mountains of the southwest, west and even Wicklow had forty-eight-hour rainfall totals that were in excess of 125mm. NUI Galway recorded 60.8mm on 17 November; its highest ever daily total for November. Cork Airport and Valentia Observatory both recorded twenty-four-hour totals in excess of 50mm on 19 November, with 51.2mm and 50.4mm respectively. Other stations recorded their highest daily totals for a November day of between ten and thirty years. So, on top of the background of very high rainfall levels, there was an exceptional period of high-intensity rainfall over a few days. This was the trigger to cause the main flood event to occur in the west and south in particular. The exceptional rainfall during this period had recurrence intervals of between one and fifteen years in most places barring Co. Galway.

In Co. Galway, the recurrence interval of this two-day rain spell varied between one and 134 years, the latter value being calculated at the NUI Galway station (Walsh, 2010).

DECEMBER RAINFALL: SUSTAINING THE FLOOD

The rainfall of November continued unabated well into December and it was only on 10 December that the synoptic situation changed and the wet spell ended and the cold spell was on its way. Much of the rainfall of December occurred in this first nine-day period. This prolonged and added to the flooding, which was still a huge problem in the west of Ireland, especially in Co. Galway and along the Shannon. Later in the month, there were general falls of rain, sleet and snow on 25, 26, 29 and 30 December (Met Éireann, 2009). In fact, over 50mm fell in some places on the 29th and 30th, generating flooding in a number of places and, in the case of parts of Co. Galway, helping to slow the gradual retreat of flood waters.

Overall, December recorded below average or near average rainfall, but the amounts were more than sufficient to slow the lowering of flood waters and to keep the threat of renewed flooding at the forefront of everyone's minds. Valentia Observatory recorded 186.4mm in what was to become their wettest year since 1866, with a total of 2,174.9mm – well above the previous high of 1,923mm recorded in 2002. This not only broke through the 2,000mm threshold but also the 2,100mm threshold (Met Éireann, 2009). Values varied from 55 per cent of normal rainfall at Malin Head to 149 per cent at Johnstown Castle in Co. Wexford (Table 3.3).

Table 3.4 Comparison of rainfall totals and % of average (1961–90) for December 2009 (after Met Éireann, 2009).

	December	
Station	*Total (mm)*	*% of average*
Belmullet	127.9	107
Casement Aerodrome	103.6	140
Cork Airport	157.6	115
Dublin Airport	71.6	n/a
Johnstown Castle	164.5	**149**
Malin Head	57.0	55
Mullingar	74.2	80
NUI Galway	114.9	95
Shannon Airport	70.4	71
Valentia Observatory	**186.4**	117

THE FLOODING: NATURAL CAUSES

This scale and intensity of the flood came about as a result of a combination of natural factors which we have looked at here; the human factors that contributed will be examined in the next three chapters, as the flooding in Co. Galway, Cork city and the River Shannon is examined in detail.

The rainfall levels recorded after the summer were staggering. There were all-time monthly records, with unprecedented summer and annual totals being recorded at various stations, as well as new long-term records at most stations over various periods. There was no real let-up in the rainfall, with the exception of early to mid-September and mid- to late October. As a result of this exceptionally wet year, the water table did not drop much during the summer and autumn. This reduced the ground's capacity to soak up water during heavy rainfall events and meant that river and stream channels were quite full even before November.

Given these circumstances, the last thing the country needed was another period of persistent heavy rain, which was unfortunately just what we got. There was simply nowhere for the huge amounts of water that were dumped on the country to go. As a result, we got flooding on a scale not seen in a very long time – at least a century in the case of the west.

One of the more unusual aspects of this was that the key rainfall, which occurred between 18 and 20 November, did so without an associated storm. The storms came afterwards. This is most unusual, as nearly all of our high rainfall events are associated with storms or depressions crossing the country in the winter half of the year. Instead, it almost felt like semi-tropical rainfall.

CONCLUSIONS

Even if there was no human involvement, there is no doubt that flooding would have occurred across many parts of Ireland as a result of this exceptionally wet year. There is also no doubt that the flooding would have been severe in many places. However, this is not the full story, as it is also clear that humans played a big role in exacerbating the scale, depth, duration and particularly the impact of the massive flooding that occurred. The remaining chapters in the flood section of this book explore the human contribution, by looking in detail at the flooding in Cork city and Co. Galway and on the River Shannon, and the factors that contributed to the impacts. Consideration is also given to the roles of various agencies, from the ESB to local and national government.

CHAPTER 4

Flooding in Co. Galway and the west of Ireland

INTRODUCTION

This chapter will focus on the flooding in Cos. Galway, Clare and Kerry. The case of Galway is significant because of both the impact and the duration of the flooding, which lasted well into the New Year in some places and had the added effect of being frozen. The scale of the flooding in Co. Galway can be gauged by the fact that only for the emergency opening of the new Galway–Dublin motorway the city would have been effectively cut off by the flood waters. Other parts of the west were also affected, but not as significantly and for as long as Galway. Many of the places that were flooded in Galway and the west of Ireland are not mentioned in this chapter, but the flooding and its impact there should not be underestimated. The chapter will not deal with flooding in east Cos. Galway and Clare associated with the Shannon Basin, as this will be dealt with later, in the chapter on the Shannon flooding.

CAUSES OF FLOODING IN CO. GALWAY AND THE WEST OF IRELAND

The flooding in Co. Galway and the west of Ireland was caused by the exceptional rainfall discussed above, the lack of any spare capacity in the river channels and other water bodies, and the exceptionally high water table. There was nowhere for the water to go, so it stayed on the surface and accumulated initially into pools in the lowest areas, gradually building up to deep floodwaters covering wide stretches of the countryside (pl. 3). The flooding persisted in some places for months afterwards because of the ongoing rainfall and the backlog of water to be drained into the river and stream channels. The water table remained very high throughout all this period, so little extra capacity for water storage became available. The flooding was topped up by the normal winter rainfall associated with the west of Ireland.

Like all parts of Ireland, flooding has always occurred in the west of the country, but higher rainfall means that this region is much more vulnerable. There is a long history of major floods in the west of Ireland, although these are not as well documented as those in other parts of Ireland. One remarkable flood in October 1816 was very similar to the 2009 event. Inundations were universal across Ireland and lakes and rivers overflowed everywhere. More pertinently, the whole countryside from Ballinasloe, Co. Galway, to Moate, Co. Westmeath, was under water. Further severe floods occurred in 1821, 1830, 1843 and 1851. 1830 was described as the wettest year ever remembered and the September rains and floods caused great damage in Galway (GBCO, 1852).

FLOODING IN CO. GALWAY

Galway County Council estimates that 3,381 square kilometres (1,300 square miles) of the county were affected by flooding at its peak. This represents 55 per cent of the land area of the county and illustrates the scale of the flooding, especially considering that much of Connemara was unaffected, so the actual percentage in the north, east and south of the county was much higher. Some sixty-five roads were affected by the flooding – some blocked for a few days, others still flooded months later. There was damage to key infrastructure such as roads, bridges and the water service plant in Ballinasloe. There were a number of calls for Galway to be given disaster status in view of the scale, duration and impact of the flooding.

The first real indication that significant flooding had started was on Wednesday 18 November when, due to heavy overnight rainfall, many parts of Co. Galway were affected by heavy flooding to such an extent that there were significant road closures. All routes into Galway city were closed, including the Dublin road. The N17 Galway–Tuam road was closed at Loughgeorge and, on the same stretch, the entrance to Claregalway was closed (pl. 4), while the junction between the N17 and the N63 Galway–Roscommon road was also blocked. Flooding occurred at Moycullen on the N59 Galway–Clifden road. The forecast suggested that very heavy rainfall was due for the next few days, so conditions could only deteriorate.

One of the many rivers to burst its banks was the Suck, which caused extensive flooding in Ballinasloe and blocked the main Galway–Dublin road on 20 November. That was if you could actually get to Ballinasloe, as the road was blocked by flood waters at Craughwell. The blocking of the main road at both Craughwell and Ballinasloe continued for a long time afterwards and was the main reason for the emergency opening of the Galway–Dublin motorway. In addition, a boil-water notice had to be issued in Ballinasloe because of concerns about flood water contaminating the town's water supply. St Michael's Square was one of the worst affected parts of Ballinasloe and all forty families had to evacuate their homes. This was despite the best efforts of the army, the local authority and civil defence personnel and the 5,000+ sandbags that were filled and deployed strategically. Worse again, the ESB had to disconnect electricity from several parts of the town because of safety risks.

A dramatic helicopter-rescue of a family of five, including an 87-year-old, occurred in the Kilbeacanty/Beagh area at 2.30am on Friday 20 November as their farmhouse was completely surrounded by flood waters, to such an extent that the Irish Coast Guard helicopter was the only mode of transport that could reach them. The house looked like it was sitting in the middle of a vast lake. Other houses in Co. Galway ended up in the same situation. The residents of Abbeyknockmoy fought a great battle to prevent serious flooding in the area and tried to alleviate the problem by digging a trench to drain away some of the flood waters. They were upset at the lack of support from the local authority, as were many rural communities. The resources of the local authority were completely insufficient to meet the demands put on them.

Driving in Galway during at least the first two weeks of the flood stretched the abilities of motorists to the limit as a result of needing to drive on flooded roads and the bewildering maze of diversions often needed to complete even relatively short journeys. The famous N17 from Galway to Tuam was blocked at Two Mile Ditch near Castlegar; this was one of many locations throughout the county where houses were flooded. The N84 Galway to Headford road was no better as there was significant flooding near Clonboo on 20 November.

Claregalway was another town in east Galway which suffered horrifically as a result of the flooding, and the local residents wondered why the local authority had not cleared the drains prior to the River Clare bursting its banks (pl. 5), swamping parts of the town and blocking the main N17 northbound.

Public transport was radically affected, as bus services out of Galway were suspended as well as rail services because of the flooding around the city. The closest the buses got was Loughrea and the closest the trains got was Athlone, further emphasising the isolation of Galway and the scale of the flooding on the 20th. Schools were closed through the east of the county, and in any case children would have been unable to get to school as a result of the extensive flood waters.

Remarkably, Galway city itself was barely affected by the flooding directly, but this was more than countered by the fact that it was surrounded by flood waters for over a week. The city remained pretty much flood-free, as all the sluices on the Salmon Weir in the city were opened in order to alleviate the pressure from water upstream of the city. The sight and noise of the water as it thundered through the city was spectacular as well as frightening.

By 23 November, there had been only marginal improvements in the flooding, and any belief that the flood waters would disappear as quickly as they had arrived was rapidly dispelled. Large parts of Co. Galway to the north, east and south were still seriously flooded, with severe disruption to traffic. The main Galway–Dublin road at Craughwell had re-opened, but was still blocked at Ballinasloe, where water levels dropped by 23cm but not enough to alleviate flooding in the town. The Derrymullen area was still badly flooded and some of the evacuated houses remained empty for weeks afterwards. The motorway was still in use as there was no other route around Ballinasloe that could be easily used and did not involve very extensive diversions. Bus services on the Tuam road were affected by diversions at Claregalway, where the road was still blocked by flood waters (pl. 6).

Ballinasloe was still a disaster zone and over one hundred families had been forced out of their homes because of flood waters. These evacuees were mostly from new housing estates that had been built on floodplains and were critically affected by the flooding. Many people had been unable to secure flood insurance prior to this flood event because of previous flooding and there were calls for emergency aid as a result to those who had no flood cover. A precautionary boil-water notice was still in place in Ballinasloe. There was a similar scenario in Claregalway, with some homes evacuated and another twenty-five families on standby. Across the county, many families self-evacuated their homes due to the flooding.

The complex traffic restrictions were still in place and it was particularly difficult to move northwards from the city. The N84 Galway–Headford road was the only route northwards still operational despite heavy but passable flooding, yet because of the heavy usage there were four-hour delays at times. The N63 Galway–Roscommon road was still impassable between Loughgeorge and Annagh Cross, further limiting people's options.

One of the big concerns that emerged was the number of rural-dwellers marooned, along with huge numbers of stock. According to the Irish

Farmers Association (IFA), there was potential for serious environmental and health problems because of the flooding of slatted houses and overflowing septic tanks (Siggins, 2009a). Galway County Council was placed in an impossible position with the number of requests for help, and in many cases even if they could have pumped water there was nowhere for it to go. As a result of being stretched to the limit, they had to make the decision not to pump out private homes as all staff were needed just to keep essential services going. On 25 November, the main Galway–Dublin road re-opened at Ballinasloe and the temporary opening of the new motorway was ended.

By 3 December, conditions had improved but there were still significant flood problems in many areas. Hundreds of homes were still flooded in the west, especially in south and east Galway, where 6,000 householders faced a two-month boil notice as the council attempted to flush out contaminated water in the public water system. Water levels were still up to a metre deep in some homes in Ballinasloe and the residents were still unable to return to start the process of cleaning-up. Indeed, the whole countryside was still sodden in the extreme and water channels were full to the brim and only just starting to drain away a lot of the excess water on the land and in the soil. There was some sense that the situation was resolving itself, but there were still great worries about any period of sustained rainfall that might result in serious flooding re-emerging.

South Galway and Gort: inundated once again

In many respects, what happened in Gort and south Galway is a micro-cosm of what happened across the country. In some ways, however, it was different because of its long history of flooding. The town of Gort is sadly synonymous with regular and repeated flooding, and the lack of progress by the authorities in dealing with the problem. This resulted in the town being badly affected again by the flooding, much to the exasperation and despair of the townsfolk. Numerous reports have been produced on the problem in Gort, yet little improvement has been achieved. It is hoped by many in the affected area that the recent flood will finally force the authorities to deal with the flooding in a systematic and comprehensive way. Anything less would be a travesty.

Prior to 2009, the last major flood to affect Gort was in 1995, but there was smaller scale flooding on a regular basis throughout the interim. One of many reports on the 1995 flooding stated that it would take IR£22 million to resolve the problem with drainage works. The report, prepared by Jennings O'Donovan, also pointed out that this would have serious environmental and economic implications. There are areas protected under the EU Habitats Directive that would have been affected by the drainage works and this was seen as a major reason to oppose this solution to the problem. As a result of the concerns raised by the report, the scheme never went ahead. The one obvious question that springs to mind after the recent major flooding is how much environmental and economic damage has been done, and it would be very interesting to compare this to what would have happened if the drainage scheme had gone ahead. Until a long-term solution is found, even more damage will be done if and when flooding recurs.

In the flood of 20 November, the main road out of Gort was blocked by water that stretched virtually all the way to Kiltartan on the Galway–Limerick road. This could only be described as a lake that was impassable by all vehicles. As a result, the diversion on the N18 to Ennis and Limerick started at Oranmore, as there was severe flooding at Ardrahan and Labane, which were just about passable, and were only recommended for local traffic. Similarly, the diversions to Gort on the N66 started in Loughrea such was the scale of the flooding. The major Galway–Limerick road was impassable and remained so for a long time.

There was extensive flooding in Gort itself on the 20th, and this stayed for several days. Crowe Street was once again one of the main flooded streets. Only for the efforts of a number of local business people in procuring pumps and pumping vast quantities of water out of this part of the town, the situation would have been much worse. Businesses in Gort were badly hit by a combination of the town being effectively isolated by the flood waters for several days, with complicated diversions in place, and the actual damage caused to many premises in the town.

Conditions on the N18 did not improve for several weeks, and by 23 November, the N18 was still impassable at Kiltartan and Labane but was passable even though heavily flooded between Clarinbridge and Kilcolgan. On 2 December, conditions were not much better, as at Ardrahan, Kilamoran and Ballinastague near Labane, north of Gort, flood problems persisted due to swollen underground and over-ground streams running through turloughs trying to make their way to the sea at Kinvara. It is interesting to note that many of the flooded locations were not on floodplains, indicating the clear need for some sort of a drainage scheme. Instead, much of the flood waters came from the runoff from sodden land draining into subterranean waterways. The N18 reopened at Kiltartan on 1 December, but there was still one local diversion on it at Labane and this diversion was still in place into the first week of February.

By 25 November, the N18 Galway–Limerick route was still blocked at Labane and homeowners in Ardrahan, Gort and Portumna were still building temporary defences, despite the bad weather. Kiltartan and Skehan were still badly affected, with over 2m of water in Kiltartan Church. In addition, farmers in Ardrahan were on high alert for fear of further flooding and the villages of Kilamoran and Ballyglass were worried about rising flood waters that were a threat to a number of houses. There was still flooding on some of the minor roads in south Co. Galway, and some of these roads were still impassable weeks later. Flooding was starting to recede in some of the worst hit areas south of Gort, but this trend needed to continue for some time to alleviate the situation.

FLOODING IN CO. CLARE

Ennis was affected by flooding to such an extent that the Cappahard Lodge Resident Psychiatric Unit was evacuated for fear of backed up sewerage and flood waters on Thursday 19 November. Twenty-eight patients were evacuated. Once again, the railway line between Ennis and Limerick was flooded, resulting in bus transfers for the passengers on 20 November. By 25 November, there was real concern about potential serious flooding in the west of the county as the Ennis basin, on the western side of the county, was full and was likely to flood with any additional rainfall over the next few days. Even by 2 December, minor roads in north Co. Clare were still flooded and in some cases impassable, especially where there were hollows.

FLOODING IN CO. KERRY

The biggest impact of the flooding in Co. Kerry was to block some key roads. There was severe flooding at Moll's Gap, which was estimated to be around 1.3m deep. The Killarney–Tralee road and the Farranfore–Castleisland road were also extensively flooded in places. The N71 Ring of Kerry was closed because of flood waters and rising lake levels. The Lake Hotel in Killarney had to be evacuated as the lake rose to its highest level in living memory. Parts of the hotel were flooded, including the laundry and other service rooms, and desperate efforts were made to keep the waters from getting into the main part of the hotel, using sandbags and pumps even though the waters of the lake had reached the outside wall of the dining room on 20 November.

SENSE OF COMMUNITY

The sense of community was awoken in many people throughout the county of Galway who helped those who were in distress, doing everything from filling sandbags to providing help in evacuations to providing accommodation. One Ballinasloe resident, Geraldine White, even set up a website called www.offerstohelp.com, where people and businesses could pledge services and items that might be needed by those affected by the flooding, including services, accommodation and furniture.

CONCLUSIONS

Flooding in Cos. Galway, Clare and Kerry was directly due to the intense rainfall and already high water levels. The scale of the flooding was enormous and hundreds of homes suffered serious flooding, not to mention the traffic chaos and the damage to infrastructure. Some of the homes that were flooded were in new or relatively new housing estates built on floodplains, exposing people to exactly this type of event. Flood problems were still occurring in pockets well into 2010, demonstrating just how severe this event was.

CHAPTER 5

Flooding in Cork city and county

INTRODUCTION

One of the most disastrous and dramatic events of the whole flood saga was the flooding of Cork city (pl. 7). The city experienced flooding on a scale not seen for several decades and this flood ranks as one of the severest on record for Cork. Miraculously, no one lost their life, but that was more by good fortune rather than any other factor. Unlike other major floods to have affected the city over the centuries, however, this flood had a considerable human component and this and the contribution they made to the flood event will be considered below.

THE VULNERABILITY OF CORK CITY TO FLOODING

Cork has a long history of flood events dating back to the foundation of the city. This is a reflection of the vulnerability of the city to this type of event. There are many factors that enhance this vulnerability. Firstly, the city was founded on two inter-tidal marshy islands in the River Lee, the site being chosen for defensive and shipping purposes. As the city expanded, a further eleven islands were reclaimed and connected to the city, but all were vulnerable to flooding. This means that the central city area is very low in comparison to mean sea-level and is vulnerable to very high tides and storm surges. In addition, the central city area is flat and sits in a narrow valley with very steep sides, all conducive to turning this valley into a basin full of water. Over time, even the steep sides of the valley have been urbanised, thus increasing the speed at which water pours down the hills into the central city area (Hickey, 1990).

As well as being vulnerable from the tidal side, the city is also vulnerable from the river side, and high and/or sustained rainfall is capable of generating floods of its own accord, irrespective of the tidal situation. Usually, the worst floods are some combination of the two. The construction of the Carrigadrohid Dam and the Inniscarra Dam on the Lee in the 1950s was seen by many at the time as being a significant step in reducing the likelihood of flooding in the city. However, the dams were not set up for flood control but for hydro-electric power generation (Hickey, 1990).

As a result of these factors, the city experienced flooding throughout its history (Tyrrell and Hickey, 1992). The central city area between the north and south channels, comprising the main shopping and commercial district of the city, is flooded on an almost yearly basis. Thoroughfares like Oliver Plunkett Street, Patrick Street and the Grand Parade are among the most flooded streets in the city. Oliver Plunkett Street was mentioned in 102 out of 277 reports of flooding between 1841 and 1988 and indirectly implied dozens more times. In the latter two cases, these streets were originally water channels and had docks for ships to tie up at and were paved over and converted into streets as they became too small for their original maritime purpose (Hickey, 2005). Flooding also extended up the Lee Valley inland from the city, and many riverside areas including the Lee Fields were left undeveloped because of the frequency of the flooding during the winter half of the year. These floodplain areas were used for parks and sports grounds, such as the Mardyke, which included University College Cork's sports grounds, among others.

SOME MAJOR CORK CITY FLOODS

One of the earliest recorded floods in Cork city was in 1633, when the old North and South Gate Bridges and the castles on them were washed away (Tuckey, 1837). The flood of 17 January 1789 was also of an exceptional nature and there was one fatality – a man named Noah, ironically. Water depths over a vast area of the city were between 1m and 2m, and this was one of the rarest of flood events in that a contributing factor was significant snowmelt along with very heavy rainfall. Much damage was done to the city as a result of the flood.

The 1789 flood was probably only exceeded in severity by the flood of 1–2 November 1853. This notorious flood claimed the lives of seven or eight people at least, including a number who were on the old St Patrick's Bridge, which partly collapsed as a result of the pressure of the flood waters. The flood was caused by a prolonged wet spell, running for several weeks with no let-up. Flood depths of between 2m and 3m occurred throughout much of the flat part of the city up the Lee Valley, and the damage and destruction was incredible. It was some time before the city started to recover (Smith, 1839).

Another major flood occurred on 26–27 September 1875 and was associated with both high rainfall and tidal flooding. Similarly, from 17 to 19 December 1945, the twice daily high tides brought repeated flooding.

More recently, flooding occurred on 21 November 2002 (Hickey, 2005). Two floods of particular interest are firstly the 4 December 1960 flood, which affected low-lying ground adjacent to the river, along the Carrigrohane and the Lee roads, where nearly 1m of water was noted, and secondly, the 25 October 1988 flood, which affected the flat part of the city, where over 50cm of water accumulated, including the Bridewell Police Station, Winthrop Street and Cornmarket Street (Hickey, 1990).

These floods are of interest as they were both partly caused by the release of water from the Inniscarra Dam. In the case of the 1960 event, the *Cork Examiner* noted that 'the ESB dam at Inniscarra has been "spilling" since the middle of last month because of the bad weather, and the weekend rain caused it to do so again, thus contributing to the flooding of many parts of the Lee Valley' (anon., 1960, p. 8). The River Lee discharge at the Waterworks Weir was 215 cubic metres of water per second. For the 1988 flood, a similar report in the *Cork Examiner* noted that 'a release of water from the ESB dam at Inniscarra created further problems downriver last night as the Lee swelled its banks, flooding already saturated lands and closing off roads' (Barker and Cassidy, 1988, p. 2). The discharge at the Waterworks Weir this time was 181 cubic metres per second. By no means were these the only occasions when the dam played a role in the flooding of the Lee Valley and the city. Martin (2009) notes a similar occurrence for the flood of 14 December 1964.

THE MAIN FLOOD EVENT OF 2009

The heavy rainfall throughout the summer and the record rainfall of November 2009 were huge factors in the flooding of Cork city, by maintaining high water levels in river channels and in the ground. Indirectly, they also forced the ESB to release around 3.8 million tonnes of water from the Inniscarra Dam for fear of the dam being over-topped and then the possibility of dam failure. This is a staggering amount of water and according to English (2009), was three times the daily volume of the mighty Mississippi River in the USA. At its maximum, the dam was releasing 535 tonnes of water per second (vastly in excess of previous discharge records from the Waterworks Weir). This is more than four times its normal rate. At its worst, the catchment was receiving over 800 tonnes of water per second because of the heavy rainfall. The ESB claimed this was a one-in-800-year event (statistically, it should only occur to this extent every 800 years on average). Given that the dam has only been in existence for less than sixty years, it is hard to see how this figure was calculated and justified, given that previous floods of this scale have affected Cork over the past 300 years (in 1789 and 1853, for example). It must be noted that only 60 per cent of the catchment is controlled by the two dams and uncontrolled water from the Bride and Shournagh tributaries of the Lee, which are downstream of the dam, would have added to the flooding.

The release of water occurred in the early hours of Friday morning, 20 November, and swamped much of the low-lying parts of the city, the areas normally vulnerable to flooding, including all of the central city island between the north and south channels, but also areas that had never flooded before or for which there were no records of previous flood events. Some people awoke in their beds on ground floors to find flood water only a few inches away from their faces. Water depths in excess of 1m occurred throughout many parts of the city. The Mercy Hospital was surrounded by water and there was significant damage to Bachelor's Quay. The Mardyke area was worst affected, especially the Mardyke Arena. Thoroughfares badly flooded included Patrick Street, Grand Parade, South Mall, Shears Street, Washington Street, North and South Main Street, John Street, Centre Park Road, Carrigrohane and Sunday's Well Roads, the Western Road (pls 8, 9), Wandsworth Quay and the South Terrace to name but a few.

The list of damage is almost endless, but one of the most critical outcomes was that the city water-treatment plant was completely covered in water, meaning that nearly half the population of the city was without clean water, which is ironic, considering that much of the city was covered in water. Homes, businesses, the university and the Westley five-star hotel (1.3m of water), not to mention the County Hall and the brand new County Library on Carrigrohane Road, were all severely affected. Cork city came to a standstill for a week, with schools and many other facilities closed in the immediate aftermath of the flood. The total bill for damage to the city probably exceeded €100 million.

One of the positive features of the flooding was the response of all the emergency services, from the professionals right through to the volunteers, along with the army. The event produced a great sense of common purpose, and the communities pulled together to deal with the problems as they arose as best they could. One of the key tasks was to get the water supply working again for nearly half the city. This took over a week, as repairs had to wait until the water treatment plant was above the residual flood levels. In the mean time, a huge water supply system developed using tankers and bottled water in order that people could get by in the short-term. This was mostly run by the army.

THE ROLE OF THE ELECTRICITY SUPPLY BOARD (ESB)

There has been controversy since the flood about the role of the ESB and its release of water into the Lee. Some of the controversy has focused on the actual release of the water and why water levels had not been reduced in advance, given that heavy rainfall had been forecast. Water levels were reduced prior to the flood, but not by enough, as the actual rainfall vastly exceed what was predicted (90mm as opposed to 46mm). Most people accept that in the end the ESB had no choice but to release the huge amounts of water, for fear of the dam over-topping and possibly failing. Failure would have resulted in a huge wall of water travelling down the Lee Valley and into the city, and the effects would have been similar to that of the Southeast Asian earthquake and tsunami of 2004. There is no doubt that the fatality level would have been very high if that had occurred. The impact on the city would be equally devastating, with buildings and infrastructure being swept away by the force of the water.

Most of the controversy about the release of water focused on the issue of communication, especially claims and counter-claims regarding the scale of the warnings issued by the ESB, to whom they were issued, and the fact that most people were unaware of these warnings, even in the most vulnerable areas. One of the websites used to communicate the risk had an average of only 1,700 hits per day (English, 2009). The communication failure is emphasised by the miraculous escape of many people who were asleep in their ground-floor beds, oblivious to the rising waters outside and inside their bedrooms. Few businesses and University College Cork had taken any precautions as they too were oblivious to the imminent flooding. A lot of the anger directed at the ESB and local authorities related to the fact that people felt that, given even a little warning, much could have been saved that was ultimately damaged or destroyed by the flood waters and lives would not have been put at risk.

One of the main issues associated with the disaster in Cork is the perceived role of the ESB. Many people thought that, once constructed, both dams would be very significant in terms of flood-control and that they would lessen the vulnerability of the city to flooding. In reality, the running of a hydro–electric station is all about keeping water levels as high as possible, as this produces more head or pressure when the water is released and generates more electricity at a higher efficiency level.

The story is more complicated. Three ESB reports, including one from 2001, had warned that the reservoirs at Inniscarra and Carrigadrohid were too small to handle very large water volumes. During extreme weather events, the operation of the dams as flood alleviation facilities was limited, and there was a false sense of security among the population of Cork city. In addition, it was noted by the Lee Catchment Flood Risk Assessment and Management Study Hydrology Report that water flowing into the Lee from tributaries downstream of the dams, most notably the Shournagh and Bride rivers, was not taken into account when the engineers at the Inniscarra were calculating how much water should be released from the dams (Melia and Riegel, 2009).

As a result of this, and the operation of the Parteen Dam on the Shannon, the ESB will be obliged to reconsider its operational procedures in order to prevent the recurrence of flooding. This will most likely lead to a lower maximum operating level and much more rigorous assessment of the need for precautionary water level reductions in the reservoirs behind the dams. This will affect both the amount of electricity produced and the efficiency of the dams. But it is what the public demands, and a repeat of what happened at the Inniscarra Dam would be unthinkable.

THE GLUCKSMAN GALLERY FIASCO

One of the most shocking aspects of the flooding of Cork city was what happened at the Glucksman Gallery on the main campus of University College Cork (UCC). The gallery was constructed as part of Cork's tenure as European City of Culture in 2005. The gallery is built on the lowest level of the main campus (known as the Lower Grounds), beside the south channel of the river. This area was previously used by UCC for tennis courts and had a small wooden pavilion with changing rooms. There was also an extensive grassy area that had no structures on it apart from a few benches and a gravel path.

On the night of the flooding, the south channel overflowed extensively into the Lower Grounds of UCC and rapidly flooded the basement of this building and eventually started to flood the first floor, which is also the main entrance level and is well above the actual ground level. The main exhibition spaces were on the first and second floors. On the night in question, the basement contained 184 works of art of both national and international importance. Despite the valiant efforts of the staff, a large number of paintings could not be rescued from the basement as a result of the rapidly rising waters. Other properties belonging to UCC were also flooded, including the Student Village (previously a riverside field) and the Western Gateway Building (previously the Cork Greyhound Track). In its old guise, the greyhound track flooded fairly regularly, including on 11 November 1941 and 14 December 1964. More recently, it flooded in February 1990 and November 2000 (English, 2009). In all, twenty-nine buildings in UCC were affected by flooding.

The damage to the gallery was serious but not structural, but the damage to the paintings was considerable. Some works of art were completely destroyed. Many other paintings needed extensive conservation work and many are still undergoing careful repair. It will be some time before the full impact of the flooding can be accounted for in the gallery.

In many respects, what happened at the Glucksman Gallery is a clear example of the flood problems in many locations across the country. It is evident that a number of serious errors led to the disaster. It was a big mistake to build the Glucksman Gallery on what is a part of the floodplain of the south channel of the River Lee. It is hard to understand how planning permission was granted on the basis of this land area being described as 'former' floodplain, when clearly it still was, being located only a few metres from the south channel of the Lee. Previous flooding events occurred at this site in November 1892 and in November 1916, when the nearby Donovans Road Bridge was swept away by a flood. To cap it all, to store irreplaceable art works in the basement, which is the most vulnerable floor of the building, just compounds the errors. It was an accident waiting to happen.

FLOODING IN CO. CORK

Many areas throughout Co. Cork were also badly affected by severe flooding and many roads were impassable, most notably nearly all routes into Bandon, which was itself badly flooded. The Fermoy–Mallow road was inundated, while parts of Fermoy were also under water. Some of the worst affected areas were Ballingeary, Crookstown and Inchigeela in the Lee Valley, and Bantry and Clonakilty in West Cork.

CONCLUSIONS

There is no doubt that the scale and intensity of flooding in Cork city was a direct result of the excessive rainfall leading up to and during the event. The operation of the ESB dam at Inniscarra contributed significantly to the flooding that occurred and their operational procedures will have to be revised in order to prevent any future event of this nature, as will communications between the ESB and the local authorities and the ordinary people of Cork, who suffered the most. This review is currently underway and involves the modelling of dam discharge scenarios to define flood risks and extents (Roche, 2010). This needs to be carried out in the light of predicted increases in winter rainfall as the century progresses, especially as forty-eight-hour severe rainfall events will become more common and these are the ones most likely to cause flood problems (Hickey, 2008). In addition, the granting of a whole series of planning permissions for important buildings on known and active floodplains is beyond comprehension, and was a recipe for disaster. For many of these locations, it was not a question of a calculated gamble; it was a question of 'when' and not 'if'.

Interestingly, the cost of the November flood event at around €100 million and possibly as high as €145m almost exactly matches the estimated cost of flood defences which would be needed to prevent any future major flooding in the city (Kelleher, 2010). However, this amount is more than double the €50m government allocation for flood defences for the whole country in the light of the 2009 event. The demand for this money will be enormous.

CHAPTER 6

Flooding on the River Shannon and elsewhere

INTRODUCTION

This chapter is primarily concerned with flooding on the River Shannon and the important role of that river in draining nearly one third of Ireland (twelve counties including parts of east Cos. Galway and Clare). The movement of the flood waters downstream towards Limerick city is discussed, and the role of the ESB in its running of the Parteen Dam is also considered. It would be impossible to describe in detail the daily pattern of flooding due to the amount of material available, and instead the chapter focuses on key periods and incidents, without underestimating the impact of the flooding in places that are not mentioned.

CAUSES OF FLOODING ON THE RIVER SHANNON

The River Shannon is the longest river in Ireland and Britain. From its source at Shannon Pot in Co. Cavan to Limerick city, where it becomes tidal and an estuary, it is 358km (224 miles) long. In addition, the Shannon Estuary extends for a further 96km (60 miles) before reaching the Atlantic Ocean. The Shannon receives water from twelve counties and from many lakes, the three largest and most important being Lough Allen, Lough Ree and Lough Derg. It is very much a lake-dominated river system, and this is a result of the high rainfall levels and impeded drainage (Heery, 1993).

The Shannon has a long history of flood problems and this is reflective of the size of the catchment and most importantly the lack of gradient on the main channel of the river. Ireland has been described as being bowl-shaped in terms of the landscape, with most of the mountains being close to the coast, and the interior being flat and low-lying. This makes it very difficult for rivers to drain their catchment areas rapidly, especially after major rainfall events. This is one of the main reasons that Irish rivers tend to have such wide floodplains and why they regularly flood.

In terms of the actual gradient on the Shannon, the situation could not be much worse. In fact, over the 205km from Lough Allen to Lough

47

Derg, the river only drops 12m. Between Athlone and Shannonbridge, a distance of around 20km, the river only drops 35cm. This means that the Shannon has the shallowest gradient of any major river in Europe. Much of the Shannon Basin is consequently very vulnerable to flooding, especially after heavy and/or prolonged wet periods. The latter can substantially raise the water table and further reduce infiltration capacity, generating much overland flow and flooding. As a result of this, the river has very extensive callows or seasonal flood areas, which are not built on.

Numerous flood studies have been carried out on the Shannon, including one by L.E. Rydell in 1956. Rydell was a member of the US Army Corps of Engineers, which worked on flooding along the Mississippi. He was brought over by the Irish government in an attempt to see if the problem could be resolved using the experience of the Corps of Engineers. Rydell (1956) concluded that flooding would always be a problem on the Shannon as a result of the poor gradient of the river and he could not envisage any large-scale scheme that would completely resolve the problem.

Although the main channels of the River Shannon and its associated tributaries have been little altered by humans, much arterial drainage of the land which feeds these rivers has been carried out over the last 200 years in order to create vast tracts of usable farmland and to help solve local flood problems. This process is still occurring. Almost every year, however, flooding happens on some part of the Shannon catchment.

Clearly, the persistent wet summer and the exceptional November rainfall of 2009 were ideal for creating a flood along the Shannon on a scale that had not been seen for several decades. In addition, because of the huge amount of water that drained into the Shannon, the ESB was faced with the same dilemma here as at Cork: either release large amounts of water at the Parteen Dam, even though it would cause flooding down-stream, including in many riverside towns, villages and housing estates, or watch the dam being over-topped, which then raises the possibility of a dam failure and a much more catastrophic result. In the end, despite the consequences for many people downstream, they increased the output of water. As a result of the length of the Shannon, flooding moved down-stream in a systematic way, as is typical of major river systems, so people and local authorities had some warning of what was coming and could try to prepare as best they could.

PREVIOUS FLOOD EVENTS

This river has a very long record of flooding going back to entries in the medieval annals. There are mentions of floods in AD942, when it was noted that there was a great flood on the Shannon. Flooding is an annual feature of the river, but every now and then floods of exceptional severity occur. Other major floods occurred in AD1251 and in 1705, when half of the population of Limerick was affected by flooding. In November 1821, thousands of hectares of land were under water, driving many inhabitants from their dwellings. In 1822, 1828 and 1851 there were great floods along the Shannon at Limerick (GBCO, 1852).

THE MAIN FLOOD EVENT

The initial flooding on the River Shannon occurred on 20 November. Carrick-on-Shannon, Co. Leitrim, was under threat of being cut off by rising flood waters as the town was already affected by some flooding (pl. 10). Roads were closed in parts of Co. Leitrim, including the main Carrick-on-Shannon–Dublin road, the N4. Leitrim village was also badly affected by rising waters. This was considered by some to be the worst flooding in the town since 1973. Some of the developments on the riverside of Carrick-on-Shannon were among the worst hit areas, emphasising the scale of inappropriate development on floodplains. Many saw the developments in Carrick-on-Shannon as a classic example of indiscriminate and revenue-earning planning permissions being granted on or near flood plains, and there was huge anger in the town as a result.

There were also extensive road closures in parts of Co. Roscommon, with twelve roads being affected. The most important of these were the N60 Ballymoe road, Laragh and Enfield to Ballintubber and the Milltown–Castlerea road, while areas around Boyle were also flooded (McDonagh, 2009). In Limerick city, the giant metal Christmas tree slipped its moorings as it was being tugged into position in the middle of the river, such was the force of the water. It eventually crashed into Shannon Bridge, where it had to be righted and dealt with as it was now at a 45-degree angle due to the force of the water. The bridge had to be closed as a safety precaution until the tree was moved to safety.

On 23 November, rising levels of water on the Shannon affected the N65 between south Galway and Tipperary, the bridge at Portumna being impassable. In addition, parts of east Co. Clare were under threat because of rising levels in the Shannon, as the flood water moved southwards to areas such as O'Briensbridge, Clonlara, Westbury and Shannon Banks. The ESB warned that more water would have to be released from the Parteen Dam, which would flood many rural areas as well as Limerick city, downstream of the dam.

By 25 November, flooding on the Shannon affected the entire river basin area, from Leitrim and Roscommon down to Clare and Limerick, and was probably at its peak. The Shannon had flooded significantly up to Athlone, Co. Westmeath. Thirty personnel from the army were needed for emergency work as water levels rose in Athlone. Focus was now turned to the Parteen Dam, which is part of a hydro-electric station operated by the ESB. A similar scenario to what happened in Cork started to emerge, with the rate of discharge at the dam crucial to determining the extent, severity and duration of flooding downstream. Water was quickly released and more would have to be released as the flood progressed, even though water levels were at record heights.

Even without increased discharge, higher flood levels occurred in the lower Shannon area, causing problems for south-eastern Co. Clare. There was concern that the ESB might have to release more water at the Ardnacrusha Dam as well as at Parteen. The increased release of water at Parteen became critical, and by 27 November there was huge concern among local authorities about how they would cope with the expected record water levels. Even worse, further serious flooding looked likely because high tides were due the following week, possibly raising already record flood levels by a further 1m. This was especially the case in Cos. Limerick and Clare. The minister for the environment, John Gormley, had to assure locals that the dam at Ardnacrusha could withstand the pressure. Householders were on high alert across the lower Shannon, but especially at Montpelier and Castleconnel, as the ESB released even more water and raised levels by a further 6cm. This set new records for water levels and the flooding downstream worsened considerably (Hayes, 2009). Vast areas in the Shannon Basin were flooded, as were numerous waterside towns and villages.

The changeover to December finally saw a start in the decline in water levels in the Shannon Basin, much to the relief of everybody, but given the scale of the flooding and the amount of water still to be drained off the land, this turned out to be a slow process, running over a couple of weeks. By 2 December, farmers in flooded areas of Co. Offaly saw water levels continuing to drop, but thousands of hectares of farmland were still under water, including many farmhouses and other buildings. The river at Shannon Harbour saw a steady fall in water over the next two days.

However, there was still over 1m of water in the flooded and evacuated houses. The ESB said that the water level at Lough Derg had dropped another 5cm and that downstream of the Parteen Dam a drop of 6cm or 7cm occurred as the day progressed.

Even as late as 3 December, hundreds of homes were still flooded despite the fact that the Shannon Basin between Ballinasloe and Portumna had seen a steady fall in water over the previous two days. Water levels were still over 1m in some homes in Ballinasloe, Athlone and Shannon Harbour, Co. Offaly. Many people affected by the flooding were able to return to their homes by Christmas, but only to face into a major clean-up operation.

FLOODING IN CO. TIPPERARY

Once again, Clonmel faced into a severe episode of flooding along the quays and the surrounding areas. Despite the fact that the Clonmel Flood Relief Scheme was over half way completed, enough of the works had not been finished to protect this long-suffering town. Flooding of the quays took weeks to finally drain away from the time the flooding started on 19 November. Some twenty people voluntarily left their homes at the peak of the flooding for fear of further inundation, and once again the army was deployed to help deal with the situation.

FLOODING IN THE SOUTHEAST (COS. WEXFORD, WATERFORD, KILKENNY AND CARLOW)

Although well away from the main flood area, there was flooding in many other parts of the country. In Wexford, the Slaney burst its banks at Enniscorthy. Flooding was a major problem near Waterford city and across the county, and the Waterford–Tramore road was badly flooded, as were a number of roads in the west of the county on 19 November.

There was also some flooding in parts of Co. Kilkenny on the 20th. There was serious flooding in Thomastown and there was nearly 1m of water in Inistioge and also some flooding in Graiguenamanagh. The R700 through the town was closed as a result. Other roads were closed in Kilkenny, including the Gowran–Bennetsbridge road and the N9 between Kilkenny and Waterford. Kilkenny County Council issued over 5,000 sandbags to people in the most vulnerable locations. Thankfully, as the Kilkenny city flood protection scheme was completed, there was no flooding in this once vulnerable city. There was extensive flooding on a number of streets in the centre of Carlow and a boat was used to move people about. Leighlinbridge was also affected by some flooding on 20 November, and more was expected as the Barrow had not yet reached its peak.

FLOODING IN THE EAST (COS. WICKLOW AND DUBLIN)

There was even a landslide as a result of the heavy and sustained rainfall and the railway line between Wicklow and Gorey was blocked, meaning that bus transfers were necessary for the rail passengers between the two. Flood levels along the River Liffey declined slowly but the Poulaphouca Reservoir still had high water levels.

On 25 November, during a strong gale, the roof of a block of apartments in Dublin was blown off and landed in a nearby field, and the roofs of two other apartment blocks in the same complex were lifted due to the high winds. Luckily, no one was hurt in the incident (anon., 2009).

FLOODING IN NORTHERN IRELAND

Although the north of the country got almost as much heavy rain as elsewhere, this was not quite enough to cause major flooding on a scale seen in the west of Ireland. However, there was flooding in Cos. Fermanagh and Tyrone, and in particular in Enniskillen, where the main road to Dublin was closed and where an extensive sandbagging job was undertaken. Derrygonnelly, Derrylin, Lisnaskea, Church Hill and Wattle Bridge were also flooded. Lower Lough Erne reached its highest water level in fifty years, following thirty-two consecutive days of rainfall.

CONCLUSIONS

The flooding on the River Shannon was very extensive and relatively prolonged due to the level of saturation of the whole Shannon Basin and the movement of the flood water downstream over more than a week. The role of the ESB at the Parteen Dam in controlling discharges was critical in determining the extent of flooding in the lower catchment, down as far as Limerick city. The impact of the heavy November rainfall on top of what had gone before can be seen by the occurrence of flooding from parts of Northern Ireland right the way down to Cos. Wexford and Kerry. Few counties escaped serious flooding as a result of this weather crisis.

CHAPTER 7

The aftermath

INTRODUCTION

This chapter is concerned with the aftermath of the flooding. Clearly, the first major concern is the impact of the event, particularly from a financial perspective. Secondly, there is the consideration of the political response associated with the flooding. Finally, there is some discussion of the political fall-out of the flooding, in particular in relation to development and the planning process.

ECONOMIC IMPACT

The scale of devastation associated with the flooding was enormous and the true amount of damage will never be known for a number of reasons. This includes damage that is impossible to assess financially, as well as non-insurance and under insurance. The biggest impact was the effect on people who watched their homes being damaged and virtually torn apart by flood waters. The pain that this inflicts on people should not be underestimated. For many too, even if all the damage is put right there will be fears about flooding occurring again.

The floods generated every possible type of damage you can think of, from private homes to businesses to key infrastructure including roads. The insurance company figures below give at least a baseline as regards insured damage, but the real figure is much higher; a ball-park amount of €500m for the flooding alone would not be excessive. This is roughly twice the estimate of insurable damage claims by the main insurance companies in Ireland.

The Irish Insurance Federation (IIF) stated that the cost of claims as a result of the flooding was €244m by the end of February 2010. The worst county was Cork, with €141 million. This figure is highest because flooding affected a major urban centre with a high density of flooded properties and facilities. Co. Galway came next with €23 million; this figure is lower because it mostly comes from towns and villages and a vast rural area. Luckily, Galway city was pretty much unaffected by the flooding, otherwise this figure would have been much higher. The third worst affected county was Clare, with €16 million worth of claims, again mostly coming from towns like Ennis, as well as villages and rural areas. The remaining €64 million worth of claims were spread out over the rest of the country, covering at least twelve counties.

These are remarkably high and unprecedented claims for Ireland and we need to put them into context. The claims for flooding alone amount to 27 per cent of the insurance industry income in a year and 31,000 customers had made claims up to the end of February 2010. Some of the companies published estimates of the cost to their business, including one of the market leaders, Aviva. This insurance company blamed floods in the south and freezing weather at the end of 2009 for a 27 per cent fall in profits to €72 million. The company indicated that for the last three months of 2009, which included the floods and the start of the cold spell, there were €88.6 million in weather-related claims in the Republic of Ireland and probably some additional small amounts for Northern Ireland. The eventual total for weather-related claims reached €100 million. FBD stated that the flooding and freezing of November and December cost it €13.5 million, and again this figure is mostly flood-related. The company profits declined from €85.8 million in 2008 to €28.9 million in 2009, partly due to weather-related claims.

One of the downsides of the flooding and the cold spell claims is that the Irish Brokers Association warned that homeowners could face an increase of as much as 20 per cent in their insurance premiums as a result of the bad weather. Even these figures are likely to be underestimates for the flood damage, as claims were still coming in well into 2010. One of the remarkable aspects of the flooding and cold spell is that there are no figures on lost school days, as the schools are not required to inform the Department of Education.

POLITICAL RESPONSE

There is no doubt that the real heroes of the flooding were the people on the ground who helped in whatever way they could during the floods and the cold spell. This ranged from local authority workers, including those who called off their day of strike action to help deal with the unfolding disaster, all the local authorities who carried out enormous work under intense pressure, the Civil Defence, the Red Cross and a vast array of other voluntary groups who train for such emergencies, the army, who provided crucial help in the most desperate areas, and the ordinary people on the street who chipped in wherever and however they could.

The response from central government in contrast is nothing to be proud of and in many respects is reminiscent of the response to Hurricane Katrina by George Bush Jr, who had to be convinced that a major disaster had actually taken place. New Orleans is still trying to rebuild, years later. The government has a committee just like in the USA where the FEMA (Federal Emergency Management Agency) has responsibility for the co-ordination of responses to major disasters. Incidentally, its chairperson resigned as a result of its poor response to Hurricane Katrina. In Ireland's case, we have the Emergency Response Committee, but it seemed little involved if at all in the initial stage of the flooding, even though it was obvious pretty quickly that this was flooding on a scale unseen in Ireland for at least a generation.

On 23 November, the Taoiseach Brian Cowen stated that flood protection schemes would remain a priority for the government. There was some improvement on 25 November when the government announced an initial €10 million towards a Humanitarian Assistance Fund. There was significant consternation when it was announced that the Humanitarian Assistance Fund would be means-tested and that the scheme would run for many months, given the time it would take for many houses to be repaired. If it is means-tested, then some families will not get anything from the government, despite being in significant difficulty as a result of the flooding. It was also stated that an additional €2 million would be allocated to farmers affected by the flooding. Farming organisations were very angry and pointed to the fact that the dredging of rivers and other water channels had been discontinued for many years because of cost and environmental considerations and they blamed some of the flooding on this. They were also angry about the focus on protecting urban areas, usually to the detriment of rural areas, and the pittance being offered by the government as compensation.

When we compare these amounts to the value of claims, not to mention uninsured damage, we can see that this was little more than a token effort despite our current economic crisis and given the billions being pumped into the banks. The government also started making enquiries about additional European funding, but little progress was made initially.

In the meantime, the hurt and anger of so many people was evident. Some idea of the political fall-out could be seen by the comments of MEP Mary Harkin, who stated that there was 'no question of proceeding with a multi-billion rail project like the Dublin metro' while people were still trying to cope with 'either inferior water and sewerage infrastructure or totally inadequate flood-protection measures' (Siggins, 2009a, p. 6).

The government finally applied to the EU for disaster relief funding in 2010, but this fund is only available for disasters where the estimated cost is in excess of €1 billion. The total damage done by the flooding was at least half of this amount and when the cold spell damage was added, the total was close to that crucial threshold. Local authorities were asked to assess the damage in their area in advance of the government's submission to the EU at the end of January 2010. The government looked for about €50 million for remedial flood works, which it saw as the main priority. However none of this money will be available to private individuals or businesses (McGreevy, 2010b).

Even the attempts to secure money from the EU did not go smoothly, as a dispute arose with the EU with regard to how much the government could claim under the EU Solidarity Fund. The Irish Department of Finance said it could only claim €6 million, whereas the EU Commission stated that Ireland could claim €12.5 million, as the Commission sees the flooding of November as an extraordinary regional disaster. In addition, Ireland is entitled to a further €12.5 million under EU Structural Funds. This would not be 'new money', as such, but would come from monies that have not yet been drawn down by Ireland from this fund, which runs from 2007 to 2013 and has an overall budget of €900 million over this period for Ireland.

THE RESPONSE OF VOLUNTARY GROUPS

Many organisations in Ireland responded in a more systematic way to the flooding and its impact on their fellow citizens. In many respects, they were doing some of the work that the central government should have been doing. It must also be noted and acknowledged that many individuals and companies gave very generously, from money to furniture to services. For example, by 3 December, St Vincent de Paul and the Red Cross had been overwhelmed by their public appeal for much-needed goods like furniture (Siggins, 2009b).

The Community Foundation for Ireland set up a Flood Recovery Fund with an initial €50,000 and looked for donations. Their focus was to help other voluntary groups who work with the most vulnerable, and these included the elderly, low-income families with young children and people with disabilities. They used their website (www.foundation.ie) as a way of generating additional donations on top of the initial fund. Many other organisations got involved with fund-raising, including the Muslim community in Ireland, which set up a Flood Relief Committee to raise funds to help victims through the Irish Red Cross. Collections were taken up at mosques around the country. The Vintners' Federation of Ireland also set aside a fund to help its members whose premises and businesses were affected by the floods.

THE POLITICAL FALL-OUT OF THE FLOODING

One of the big bones of contention arising from the flooding was the planning process in relation to the building boom of the Celtic Tiger and the building of houses, business premises and other structures on flood plains. A considerable number of new and nearly new housing estates were flooded across the country, amounting to hundreds if not thousands of houses. There was huge outcry by the residents and some politicians about the scale and severity of devastation suffered by the residents of these estates and their bleak future prospects. This includes worries about future flooding and either the extremely high cost of flood insurance, which tends to soar once an initial claim has been made, or the fact that they will be unable to get flood insurance at all. No insurance company is under any obligation to insure them for future risk. In cases where the insurance companies *do* quote, the premiums will be so high as to block most people from being able to afford them.

This begs the question as to how we ended up with so many houses on flood plains and in vulnerable areas, given our comprehensive planning system. It must also be borne in mind that this was mostly a western half of Ireland flood, if a similar event had affected the east coast, there would have been many more houses and other properties in this scenario.

The issue of zoning has to be addressed. It is illegal to build houses on land that has not been zoned as 'residential', with the exception of small numbers of one-off houses in rural areas. As the 1990s and 2000s progressed, more and more land was zoned for residential development throughout the country. This was done as a response to an ever-rising population and a seemingly insatiable demand for housing. There is also no doubt that there was considerable political pressure for the zoning of land to take place. Zoned land and land that was likely to be zoned residential started to sell for exorbitant sums of money and the values carried on rising. This created a whole new group of millionaires around the country and converted relatively low value agricultural land into extremely high value development land. Millions, tens of millions and even hundreds of millions of euro were involved in transactions, depending upon the size of the area and its proximity to existing major population centres. Once land is zoned for development there is a *prima facie* acceptance by the local planning office that development is going to take place. Remarkably, councillors in Carlow County Council were recently still pushing for further land rezoning despite everything that has happened (Parson, 2010).

There seems to have been very little or no consideration given to the suitability of much of this land for development. Many of the newly zoned areas were beside rivers and streams and were on floodplains as a result, but little or no attention seems to have been given to the flood potential of these areas. A river view is a highly desirable quality, just as much as a coastal view is, but this zoning of unsuitable floodplain land for development represented the first failure (pls 11, 12).

The next issue was the role of developers. If a developer buys a newly zoned residential site at a high price or had previously purchased and speculated on its rezoning then that person will be very keen to develop it as quickly as possible to recoup some of the expenditure and use that money for the next project. As part of the application process, the developer may be required to carry out a flood risk assessment on the site. This was done more so in the 2000s than in the 1990s and is compulsory at present, in light of the new Office of Public Works (OPW) (2009) *Planning System and Flood Risk Management: Guidelines for Planning Authorities*.

Carrying out a flood risk assessment on any site can be straightforward or incredibly difficult, depending on the characteristics of the site. The main aims of the assessment are to identify whether the site is at risk of flooding and to calculate the one-in-one-hundred-years flood level at the site. The one-in-one-hundred-years level is the design level, as most buildings are not expected to last more than one hundred years. The easiest case is if the site is in an urban area and there is a nearby water gauge with sufficiently long enough annual maximum water height data (thirty years or more). The history of the site and a look at the various ordnance survey maps of the area will show whether there has been a historic flood problem at this site. The water level data will allow calculation of the flood level and the developer will then be able to design the ground level of the site, the road level and the entrance level of the buildings so as not to flood, even if a one-in-one-hundred-years event were to take place.

Where it gets difficult is if there is no real historic data, the ordnance survey maps do not provide the information and there is no nearby water level data. It is then up to the skill of the consultant to carry our measurements at the site and interpolate from the nearest water level gauge and to produce a flood risk assessment. This is much less accurate than the previous scenario and it is in the interest of the consultant to err as much as possible on the side of caution. In addition, a climate-change component needs to be built into the level as it is predicted. This can become a problem for the developer, however, because the higher the ground and entry levels, the more cost will be involved and if parts of the site are unsafe for any building, this may further reduce the profits that can be made. It is at this point that negotiations take place between the developer and the consultant to find a suitable set of levels that both can agree on. The developer can push the consultant to lower and lower levels as the developer is the one paying the consultant's fee. The developer may be more inclined to take a chance because the one-in-one-hundred-years flood level is just a statistical construct and could occur twice in one month or not for 250 years. In addition, once the estate is completed and the homes are sold, there is a likelihood that the local authority will take over the estate and will become legally liable for it, including in the case of flooding. Once the estate has been taken over in this way, there is no come-back in relation to the developer if a flood were to take place.

A climate-change component needs to be built into the levels, as it is predicted that winter rainfall will increase significantly by 2100 and that severe forty-eight-hour rainfalls similar to what happened in November 2009 will become more frequent (Hickey, 2008). These considerations all tend to push the ground and entry levels at the site higher and higher, which is not what the developer wants to hear.

There is no doubt in my mind that in some of the new estates where flooding occurred, a proper one-in-one-hundred-years flood assessment was carried out and was adhered to by the developer and the problem arose because the flood levels exceeded the one-in-one-hundred-years level. There is also no doubt that in some cases the flood assessment was tweaked to get lower levels for the developer and that this made these areas vulnerable to flooding. It is also clear that some sites had no flood assessment at all or if it was done it was either deliberately or accidentally done so badly as to be negligent in the extreme.

The third variable is the planning office in the county council. Its role is to ensure that proper and correct development takes place and to weed out proposed developments that are clearly unacceptable. One of the assessment criteria is flood risk. It is clear that the planning offices of local authorities were swamped with development proposals throughout the

Celtic Tiger period, far beyond their capacity to assess the merits or otherwise of each proposal. They were heavily reliant on the reports sent in by consultants as part of the application process, including flood risk assessments. In addition, given the value of the proposed developments, they are under huge pressure from local councillors and individuals to grant these applications as quickly as possible and with the minimum number of extra requirements. New retail and commercial developments also become sources of revenue in the form of rates for the local authority. This is especially significant in times when these authorities need all additional income available. There is no doubt that little or no critical appraisal was made of many flood risk assessments and the planning authorities were relying on the good standing of the consultant to produce a genuine flood risk assessment. This meant that some very poor flood risk assessments crept under the radar and got through when they should have been sent back for revision. This occurred because of staffing issues and the lack of professional expertise needed to appraise these assessments.

CONCLUSIONS

The best response to the flooding crises came from the ordinary people of Ireland and many voluntary agencies and local authorities. The role of central government and its response to the flood crises and, as we shall see, the cold spell, was assessed by a Houses of the Oireachtas (2010) committee. The report noted in relation to flooding the bewildering number of agencies involved in the management of waterways and the lack of a central agency to co-ordinate and take control in times of flooding. It also called for an independent expert investigation into the circumstances surrounding the flooding of Cork city. The planning system needs a radical reassessment in the light of the scale of flooding in new or nearly new housing estates and other properties. Before the flooding had time to subside in some parts of the west of Ireland, the country was plunged into another weather crisis, in this case a cold spell of exceptional severity, and there were even more unexpected events to come.

CHAPTER 8

No let–up: the cold spell

INTRODUCTION

Although the main cold spell seemed to last only from 25 December to 10 January, the cold was never really far away and there were further outbreaks of cold and snowstorms all the way through until the end of March and arguably, with low night-time temperatures, even into April and May. From 12 December, cold conditions with sub-zero temperatures set in, and they gradually got worse, so that by 18 December, minimum temperatures had reached –6°C and, by 25 December they had reached –10°C. It must be noted that the duration of the cold spell in Ireland was reduced in comparison to other countries because of the maritime nature of Ireland's climate, which tends to reduce extremes of temperature. This is down to the role of the Gulf Steam, which brings warm waters to Irish shores, ameliorating the climate, especially in winter time, despite the country's relatively high latitude.

THE CAUSES OF THE COLD SPELL

The main driver behind the cold spell was the Arctic Oscillation, as Gibbons (2010) points out. This high pressure area over the Arctic had forced severely cold air over northern Europe and parts of Asia. The system stabilised and as a result extended the cold spell by means of a blocking high pressure area over Greenland which pushed our normal milder Atlantic weather systems much further south than usual. As a result, these milder systems did not affect Ireland in their normal way. Instead of our mild wet winter with the occasional colder spell, we now got very severe cold spells.

This large area of high pressure (bringing cold, sunny weather) was firstly centred off the northwest, and later moved in an easterly direction. This had the effect of pushing a bitterly cold northeast airflow with low temperatures and high wind-chill across the country. The days are very short and the strength of the sun's rays is relatively weak because of their low angle at this time of year, and so it is much easier for a high-pressure cold system at mid- to high latitudes to be sustained over a relatively long period of time (weeks or even months). The cold easterly airflow over Ireland was accompanied by repeated bands of snow. There was thick ice on many ponds and lakes, indicating the intensity of the freeze-up.

This is not the full story, however, as there are two other possible factors at work in aiding the generation of this cold spell. The first is the ongoing El Niño weather anomaly which froze parts of northwest Europe. Only parts of the northern hemisphere were affected by the cold spell; elsewhere in Alaska and parts of northern Canada, temperatures although still cold were well above their normal winter figures. El Niño is an ocean current event off the west coast of South America, but when it occurs it is large enough to have global-scale climate implications. It is associated with a heating up of the planet, albeit with significant regional variations. Research is ongoing to see what effects El Niño has on Europe. The most recent El Niño ended in June 2010. Remarkably, despite our cold spell, January 2010 (at 0.69°C above the long-term average) was the fourth warmest globally since 1880, emphasising the regional nature of the cold spell (McGreevy, 2010c). This global trend continued into February, which was 0.72°C above the long-term average, March (0.85°C above) and April (0.73°C). The global warm trend persisted through the first half of 2010 (anon., 2010a) and it is likely that a new annual global temperature record will be set (Leake 2010).

The second factor that may have contributed to the severity of the winter is the lack of sunspots. Normally, the sunspot-cycle peaks and troughs over an eleven-year mean cycle and, as a result, it is fairly predictable. Remarkably, since the last cycle reached its lowest point in the second half of 2009, the sunspot numbers have stayed very low. This indicates that solar output is slightly reduced, as previous long-term cold spells have been associated with few or virtually no sunspots. An example is the period of sunspot activity known as the Maunder Minimum (1645–1715), when virtually no sunspots were recorded. This period coincides with the severest part of the Little Ice Age (Eddy, 1980). This impact is mostly associated with European temperature changes, it seems, as the jet stream is pushed further north, resulting in cold air from the east affecting Europe, effectively blocking our normally milder westerly airflow. This theory is new and controversial and much more research needs to be carried out to verify the idea.

The changes in sunspots will have little impact on global warming, as solar variation amounts to between 15 per cent and 20 per cent of the current warming signal at most, and solar output has been declining since

1985. One further curiosity about sunspots remains; they are colder than the surrounding normal solar areas, yet their occurrence indicates increased solar output and their absence indicates reduced solar output (Kinver, 2010). The global temperature increases recorded in the first six months of 2010 are even more remarkable in the context of decreased solar output and this emphasises the human role in climate change.

THE TEMPERATURE DECLINE OF 12–24 DECEMBER

This period was marked by a significant decline in temperatures across the country and hinted that we were going to experience our first severe winter spell since the winter of 2008–9, when we had a relatively brief period of cold weather that broke a sequence of eight previous mild winters going back to 2001. The minimum air and ground temperatures recorded show this remarkable decline, down to levels that do not occur in most winters (Table 8.1). Daytime temperatures remained relatively high for this time of year, but the contrast at night was striking, with minimum air temperatures plummeting well into sub-zero territory and even more so if the ground minimums are examined. On 18 December, air temperatures reached –6°C, which is similar to the normal minimum temperature recorded most winters. Hard frosts were now very widespread. The main decline set in from the 21st onwards, and on 22 December

Table 8.1 The decline in temperature (°C) between 12 and 24 December
(after Met Éireann, 2009).

Date	Maximum	Air minimum	Ground minimum
12th	5 to 11	–2 to 7	lowest –6
13th	1 to 9	–4 to 3	lowest –8
14th	5 to 10	–3 to 7	lowest –8
15th	7 to 10	–1 to 7	lowest –6
16th	5 to 10	–3 to 8	lowest –7
17th	4 to 9	–2 to 5	lowest –6 and below 0 at all stations
18th	0 to 7	–6 to 2	lowest –11 and below 0 at all stations
19th	3 to 8	–4 to 1	lowest –9 and below 0 at all stations
20th	0 to 6	–3 to 1	lowest –8 and below 0 at all stations bar one
21st	1 to 6	–5 to 2	lowest –10 and below 0 at all stations bar one
22nd	1 to 7	–6 to 2	lowest –11 and below 0 at all stations bar one
23rd	–1 to 5	–7 to 0	lowest –11 and below 0 at all stations
24th	–3 to 5	–9 to 0	lowest –12 and below 0 at all stations

daytime temperatures exceeded 0°C in only Cos. Donegal and Sligo, whereas Cos. Longford, Meath and Westmeath were the coldest, with –5°C even at 11pm. This matched the previous night's temperature, when snow on the ground froze over in some parts of the country. Snow and sleet were also recorded on 18, 19 and 23 December, and drifting snow became a problem on the 19th.

THE MAIN COLD SPELL OF 25 DECEMBER–10 JANUARY

This was a period of intense cold weather with substantial sub-zero temperatures during both day and night. There were severe frosts throughout the country, and ground temperatures plummeted. Snow was also frequent (Table 8.2). The two stations where minimum ground temperatures exceeded 0°C were Valentia and Malin Head, both of which are extreme west-coast stations where temperatures are ameliorated by their proximity to the Atlantic.

This was the coldest December for twenty-eight years for most of the country and the coldest month of all since February 1986. The lowest air temperature of 2009 was recorded on Christmas Day at Mullingar, with –10°C, the lowest since January 1982. The lowest grass temperature here was –13°C, the lowest since 1950 at this site. Heavy snowfalls occurred on 29 and 30 December. There were regular snowfalls, but little accumulation except on high ground as a result of the relatively high daytime temperatures, which melted low-lying snow. No-one would have guessed that these values were going to be exceeded all over Ireland in the coming weeks. The Shannon Coast Guard rescue helicopter crew said conditions of 50 knot northwest winds and driving rain occurred in Quin, Co. Clare, where they were on a rescue mission (Siggins, 2010; and see below).

The night of Wednesday 6 and Thursday 7 January was exceptionally cold and Oak Park Station on the outskirts of Carlow town recorded a temperature of –12°C. Daytime temperatures failed to reach even 0°C in Kildare and Dublin on 7 January. Between 7 and 9 January, night-time temperatures were below –10°C in most places. In Dublin, the night of 9/10 January dropped to –12°C, with several centimetres of fresh freezing snow by morning (pl. 13).

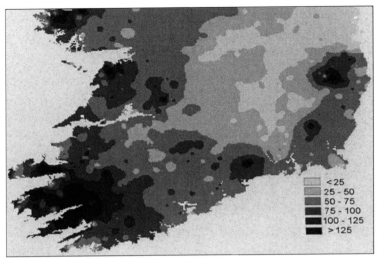

1 Two–day rainfall totals (in mm) for 18–20 November 2009 for the southern half of Ireland (image courtesy of Met Éireann).

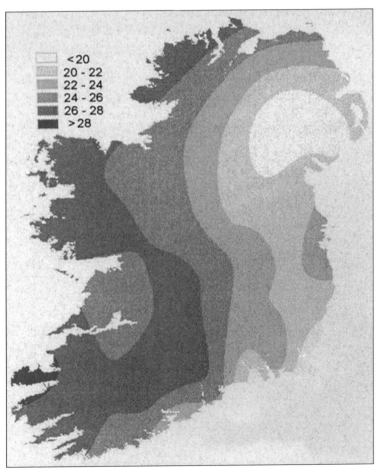

2 Total number of wet days (days with ≥1mm of rain) in November 2009 (image courtesy of Met Éireann).

3 Flooding at the confluence of the Clare and Abbert Rivers at Tonamace, Turloughmartin and Turloughcartron, Co. Galway. The place-name element 'turlough' indicates an area prone to flooding – a turlough is a seasonal lake that fills up in winter or in times of heavy rainfall and drains away in summer (image courtesy of the Office of Public Works).

4 Flooding of the River Clare at Claregalway, Co. Galway (image courtesy of the Central Fisheries Board and the Office of Public Works).

5 Montiagh, Co. Galway, isolated by flooding of the River Clare
(image courtesy of the Office of Public Works).

6 Flooding near Claregalway, Co. Galway
(image courtesy of Gary Fox/www.garyfox.ie).

7 View of Cork city, looking inland up the River Lee
(image courtesy of the *Irish Examiner*).

8 Flooding from the South Channel of the River Lee onto the Western Road, Cork
(image courtesy of Rob Fisher).

9 Flooding at the corner of the Western Road and Mardyke Street, Cork
(image courtesy of Rob Fisher).

10 Flooding in Carrick-on-Shannon, Co. Leitrim
(image courtesy of Tony Murphy/www.tonymurphy.ie).

11 Flooding at Claregalway, Co. Galway, in 2005 (compare with pl. 12, below)
(image courtesy of the Office of Public Works).

12 In Co. Galway, the Claregalway Corporate Park (top right) and the
Cuirt na hAbhain housing estate (centre left) were built between the flooding
events of 2005 (pl. 11, above) and November 2009, seen here
(image courtesy of the Office of Public Works).

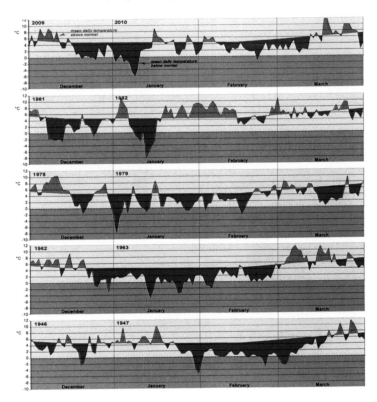

13 Luas tracks at St Stephen's Green, Dublin, during the cold spell
(image courtesy of William Murphy).

14 Comparison of recent cold spells (1946–7, 1962–3, 1978–9, 1981–2)
with that of December 2009–March 2010
(image courtesy of Met Éireann).

15 The eruption of the Eyjafjallajökull volcano
(image courtesy of Ólafur Sigurjónsson).

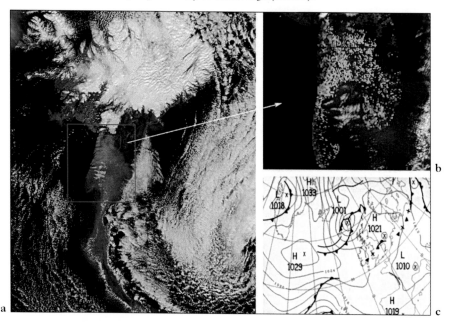

16 a) Visible satellite image at 12.50pm on 19 April 2010 of the volcanic ash plume drifting
south from Iceland; b) Enlarged area of main map, using false colour techniques to represent
the height of the ash cloud – red represents high-level ash, blue and green are low-level ash;
c) Weather analysis chart of north-western Europe at 12.00pm on 19 April 2010
(images courtesy of NASA Earth Observatory and Met Éireann).

Table 8.2 The temperature (°C) of the main cold spell between 25 December and 10 January (after Met Éireann, 2009 and 2010).

Date	Maximum	Air minimum	Ground minimum
25 December	1 to 9	−10 to −1	lowest −13 and below 0 at all stations
26 December	2 to 9	−3 to 5	lowest −7 and below 0 at all stations bar one
27 December	3 to 9	−4 to 2	lowest −8 and below 0 at all stations
28 December	1 to 8	−6 to 5	lowest −11 and below 0 at all stations bar one
29 December	2 to 6	0 to 4	lowest −1
30 December	2 to 9	0 to 4	lowest −1
31 December	2 to 7	−4 to 1	lowest −8 and below 0 at all stations
1 January	−2 to 6	−6 to 1	lowest −11 and below 0 at all stations bar one
2 January	2 to 7	−7 to 3	lowest −12 and below 0 at all stations bar one
3 January	1 to 6	−6 to 2	lowest −12 and below 0 at all stations
4 January	0 to 5	−7 to 1	lowest −12 and below 0 at all stations
5 January	0 to 6	−5 to 2	lowest −9 and below 0 at all stations bar one
6 January	0 to 5	−9 to 2	lowest −13 and below 0 at all stations
7 January	−3 to 6	−12 to −1	lowest −12 and below 0 at all stations
8 January	−4 to 6	−12 to −1	lowest −13 and below 0 at all stations
9 January	−3 to 6	−12 to −3	lowest −15 and below 0 at all stations
10 January	0 to 6	−7 to 2	lowest −7

The severity of the weather in the first ten days of January can be highlighted by the fact that it was, on average, 6°C below normal and that it was the coldest January generally since 1985 and the coldest in the Dublin area since 1963. The lowest temperature was recorded at Mount Juliet, Co. Kilkenny, on 7 January, when the air temperature measured an astonishing −16.3°C, close to an overall record low for Ireland and well below normal winter minimums. Many areas throughout the country recorded temperatures below −10°C between 7 and 10 January (Met Éireann, 2010).

Worse was experienced in Europe, with temperatures as low as −18°C in Britain, where this was the longest cold spell since 1981. Germany had temperatures as low as −20°C, with snowfall of 16cm in Berlin, where the waterways leading to the city froze over. France and many other European countries were affected by exceptionally low temperatures (Beesley, 2010). In parts of Asia, conditions were equally bad. Northern China and in particular Beijing recorded its heaviest daily snowfall since 1951 on 4 January, with its lowest temperatures for twenty years of −16°C. Heavy snow and biting cold winds also affected northern India and North Korea and South Korea.

Table 8.3 shows the minimum air temperatures recorded at each of the main meteorological stations throughout Ireland. Inland stations or those

Deluge

in the eastern half of the country tended to have the lowest minimum air temperatures in both months, with the exception of Shannon Airport in January. This pattern occurs because these areas are less influenced by our maritime position than the rest of the country and behave more like a mini-continent – colder in winter and warmer in summer. The meteorological stations on the west coast, in places such as Malin Head, Belmullet and Valentia, had the highest minimum air temperatures, showing the much stronger maritime influence in these areas. Even in March, lowest air temperatures were still negative in all stations.

Table 8.3 Lowest air temperature values for selected stations (°C), December 2009–March 2010 (Met Éireann, 2009 and 2010).

Station	Dec.	Date	Jan.	Date	Feb.	Date	Mar.	Date
Ballyhaise	−9.7	25th	−8.6	9th	−6.0	11th	−7.5	9th
Belmullet	−2.7	24th	−5.6	9th	−3.3	22nd	−3.2	11th
Carlow Oak Park	−4.9	23rd	−12.0	7th	−4.5	11th	−5.3	9th
Casement Aerodrome	−8.3	25th	−12.0	8th	−5.1	11th	−5.2	4th
Cork Airport	−2.7	21st	−8.1	2nd	−3.6	22nd	−1.7	5th
Dublin Airport	−6.6	25th & 26th	−9.5	9th	−6.6	22nd	−7.3	8th
Gurteen	−6.6	24th	−8.4	9th	−6.2	11th	−5.8	10th
Johnston Castle	−3.5	25th	−3.7	8th	−2.8	20th	−1.4	5th
Knock Airport	−6.1	25th	−5.8	10th	−3.9	21st	−4.0	8th
Malin Head	−3.0	24th	−2.9	9th	−3.3	22nd	−2.1	8th
Mullingar	−10.0	25th	−9.8	9th	−5.8	22nd	−5.9	9th
NUI Galway	−6.0	28th	−8.2	2nd	−6.5	11th	−7.1	8th
Shannon Airport	−5.6	24th & 25th	−11.0	9th	−4.8	11th	−4.6	9th
Valentia Observatory	−3.6	25th	−4.7	9th	−2.4	22nd	−3.6	5th

Comparison with other cold spells of recent times, particularly 1981/2, 1979, 1962/3 and 1947, shows that the severity was equal to these previous events, as shown in the Met Éireann diagram of air temperature from the Phoenix Park in Dublin from December to March (pl. 14). As well as the main cold spell, there was a clear extended period of below mean daily temperatures throughout most of February and the first part of March, with a re-emergence in the last few days of March. Even in April and early May, although not shown on the diagram, there were still small clusters of days with temperatures significantly below average and this shows that the meteorological conditions that caused the initial cold spell had not really gone away, it was just that their effect on Ireland was being tempered by our maritime dominated climate.

Heavy snowfalls occurred on 6 January in many places. Five centimetres fell in Cork city on the 10th, and several centimetres fell in

Co. Kerry, including Killarney, with upland areas experiencing ten hours of snow showers. Connacht was also affected by widespread snowfalls on 10 January. Additional heavy falls occurred on the 12th over wide areas in the south and on the 13th in the west.

The knock-on effect of this cold spell with high pressure was increased days with ground frosts, almost double the normal numbers for January, with between twenty-two and twenty-eight days being recorded at some meteorological stations. Record levels of sunshine were also recorded, with over one hundred hours at Valentia and Cork Airport, and there was also significantly reduced rainfall, with less than half the normal amounts recorded in some areas. Shannon Airport had just 30mm of rain in January, its lowest January value since 1963 (Met Éireann, 2010). The increase in frosts and hours of sunshine and the reduced rainfall were a product of the dominance of high pressure over the country for this extended period.

THE MAIN THAW OF 11–12 JANUARY

The main cold spell started to break from 11 January onwards, with conditions changing to rising temperatures associated with an Atlantic airflow. This milder air met a large zone of cold air centred over Scandinavia and it was the position of the boundary between these two air masses that dictated whether Ireland remained cold or not. As it turned out, the Atlantic air began to dominate slightly sooner than forecasted, and so the thaw started to set in from 11 January.

The side-effect of the rapid thaw was that the Atlantic airflow also brought heavy rain which, when combined with melting snow, caused flooding in southern areas. In Co. Kerry, the rainfall on 12 January was exceptional in nature, as a record 58.5mm fell on that day, more than twice the monthly total for Shannon Airport. There was flooding in parts of Cork city and county due to a combination of heavy rainfall and melting ice. In Cork city, some 30mm of rain fell, causing flash flooding with some buildings on the north side of the city to be flooded by up to 60cm of water. Mallow and Fermoy on the Munster Blackwater were on flood alert as a result of rapidly rising water levels in this river. Flooding was to affect many areas as the thaw really set in (see below, ch. 8).

Fresh snow fell on 12 and 13 January in the south, east and west, particularly in Cos. Wexford, Wicklow, Dublin (especially in the Enniskerry, Stepaside and Leopardstown areas), Carlow, Kilkenny, Sligo and Mayo). Met Éireann expected that this snowfall would be temporary, and not a continuation of the main cold spell. Thankfully, they were right.

THE COLD CONTINUES INTO FEBRUARY AND MARCH

The cold weather never really went very far away and Britain and the continent remained very cold throughout much of February and March (Table 9.3). In Ireland, February continued the cold spell in many respects, as it was the coldest February since 1986, with air temperatures around 2°C below normal values when compared to the 1961–90 mean. Air frosts were frequent from 7 February onwards, and a return to wintry conditions with snow occurred from the 15th onwards, with snow accumulating in northern and upland areas (Met Éireann, 2010). Lowest minimum temperatures varied from –2.3°C to –6.6°C.

It was the coldest March all over the country for between nine and fourteen years, with the exceptions of Mullingar and Valentia Observatory, where it was the coldest March for twenty-three years, since 1987. Lowest minimum temperatures varied from –1.4°C to –7.5°C, so in some parts of the country colder minimum temperatures were recorded in March than in February.

On 5 March, conditions were so bad in the Baltic Sea that high winds pushed vast quantities of ice into the Swedish Archipelago, a very unusual occurrence this far south. As a result, at least fifty ships, including a number of passenger ferries, became trapped in the ice. In the end, many ships had to be freed using ice-breakers from Sweden and Finland. The maritime administration was not happy with what had occurred, as the ships had ignored warnings about the icy conditions.

In Ireland, the snowstorm of 30/31 March caught everyone by surprise. Heavy snow, blizzard conditions, torrential rain and winds affected the northern half of the island of Ireland in particular (Met Éireann, 2010). One of the worst affected locations was the Glenshane Pass, a main route between Belfast and Derry, where snowdrifts of more than a metre blocked both ends of the pass. This had the effect of trapping over 120 vehicles and around 300 passengers, including a busload of school children. A major rescue operation had to be mounted by the PSNI, coastguard, mountain rescue and the Northern Ireland Department of the Environment. The rescued people were taken to two evacuation centres in Maghera and Dungiven in Co. Derry, but the story was not over for some, as one of the rescue centres was affected by a power cut and the people were then moved to a leisure centre near Limavady. Fortunately, everyone survived unscathed.

The blizzard also cut electricity to about 75,000 homes, north and south, while snowdrifts and fallen trees closed a large number of roads. The areas worst affected in the north (amounting to some 50,000 homes) were in Co. Derry, including in particular Coleraine, Omagh, Dungannon and Ballymena. In the south, about 10,000 homes across Cos. Cavan and Monaghan were without power; 5,000 homes were affected in Co. Donegal, as well as large numbers of homes in Tullamore, Portlaoise and Athy, and pockets across the west and east coasts. In addition, hundreds of passengers from Belfast International Airport spent the night in nearby hotels because of cancelled flights and a number of ferry crossings across the Irish Sea were also cancelled.

The event was short-lived, which is typical of a snowfall event of this nature so late in March, and by morning of 31 March, blocked roads were being cleared and Northern Ireland Electricity (NIE) and the Electricity Supply Board (ESB) started the process of repairing the electricity network and getting power to the homes affected by power cuts. Over 450 staff at NIE were drafted in to help restore power, along with assistance from several ESB crews from the south.

CONCLUSIONS

The winter of 2009/10 was the coldest in Ireland since 1962/3, with mean air temperatures over 2°C degrees lower than the long-term average for the period from 1961 to 1990. Many parts of the country saw over thirty days of snow over the winter of 2009/10, and air frosts occurred on most days, except along the coast, where they were fewer but still more frequent than in an average winter (Met Éireann, 2010). Cold weather continued through March, with some snow storms, and April, when night-time temperatures dropped below freezing and ground frosts were recorded for much of the month (Met Éireann, 2010). May was also colder than average, but June was warmer and drier than normal. July was exceptionally wet, but generally the summer of 2010 was warmer and drier than 2007, 2008 and 2009.

CHAPTER 9

Previous cold spells

INTRODUCTION

Ireland has a long history of cold spells and exceptional snowstorms, despite their infrequent nature. There are records of these weather events stretching back for hundreds of years, into the Early Christian period, as a result of the work of the Irish monks who wrote the monastic annals. Although the annals were written from the sixth to the fifteenth century, they also include details of many earlier events and provide a written record of many weather disasters that would otherwise be unrecorded. Additional records covering the medieval period come from the Table of Deaths published as part of the 1851 Census of Ireland, while later information comes from a variety of documentary and other sources (GBCO, 1852). The information is arranged below in chronological order, starting with the oldest cold spells or significant snowstorms; those that came to notice because of their exceptional severity, duration and geographical distribution, and also their impact. As we get closer to modern times, more and more events were recorded, particularly from c. 1750 onwards. The start and end dates of each climate phase are very much open to debate and depend on the authors consulted. The chronology uses the most common dates. It must also be noted that some authors dispute even the existence of these climate phases, but discussion of this is not necessary here.

THE PRE-ROMAN PERIOD, UP TO 500BC

Remarkably, three references to early snow events survive for this period, as recorded in the Irish monastic annals, however these records were written down nearly a thousand years after they are said to have taken place, so they are enigmatic and questionable as a result, particularly when the descriptions are read:

> 1023BC There was this year a huge snow with a vinous taste, from which *Oill-finsneachta* (wine snow) was so called (Britton, 1937).

665BC A great snow blackened the grass and was called 'The Wine Snow' (GBCO, 1852).

538BC It rained snow continually this year. A great snow with the taste of wine fell (GBCO, 1852).

The first two records (and even possibly all three) are likely to refer to the same event, but it could have been on any of the dates given. This difficulty is something that Britton (1937) drew attention to. The records refer to unusual and initially puzzling phenomena; the blackening of the grass and the taste of wine in the earlier event, and year-round snow with a wine taste also. The most likely cause of these two events is volcanic eruptions and there are three possible eruptions in Iceland that cover this period, although none exactly fits. The eruptions were at Fremrinámur in 800BC±300, Peistareykjarbunga in 750BC±100 and Hverfjall, around 500BC. The latter two are very close to the dates of the recorded events (Simkin, 2002). The downturn in temperature and increases in snow and ice is a product of large volcanic eruptions, as they reduce the amount of incoming solar radiation reaching the ground. This effect is short-lived, however, and rarely lasts more than three years. The blackening of grass and the taste of wine are probably due to the acidic nature of the snow, which would contain volcanic aerosols and other acidic particles as well. The role of volcanoes in influencing Ireland will be discussed in detail below, in the context of the 2010 volcanic ash event.

THE ROMAN PERIOD, 500BC–AD450

For much of the period from *c*.500BC until it fell apart just prior to AD500, the Roman Empire enjoyed generally favourable weather conditions. Warm weather allowed grapes and olives to be grown further north, and good rains allowed the Romans to buy abundant crops of grain from across the Mediterranean and in North Africa. Temperatures were probably similar to those of the mid-twentieth century. This long, mild spell peaked in around AD200, after which temperatures gradually declined. Around AD300, a sharper downturn began, intensifying from AD400 to 450. This change in weather patterns went hand-in-hand with a dramatic drop in temperatures and significant alterations in rainfall and the return of long, snowy winters, which north-eastern Europe had been spared for hundreds of years (Hickey, 2008).

Very little information for Ireland survives for this time, partly because the Romans never got as far as Ireland and the first Irish monasteries were only just being founded towards the very end of this period. As a result, there is only one reference to snow over this period and this may partly reflect the improved climatic conditions, at least until AD300.

AD436 A huge snow (Britton, 1937).

THE DARK AGES, AD450–800

This period is associated with lower temperatures than previously and with frequent late and early frosts that had devastating impacts across Europe. This downturn in climate was due to a number of possible causes, including massive volcanic eruptions in AD535 and possibly again in AD540, the most likely culprit being Krakatoa near Java in Indonesia. Michael the Syrian, writing in the year AD536, noted that the sun became dark and its darkness lasted for eighteen months (Chabot, 1899–1924). Each day it shone for about four hours, and still the light was only a feeble shadow and the fruits did not ripen and the wine tasted like sour grapes. This was a much bigger eruption of Krakatoa than the 1883 event and caused a significant temperature drop for years after, possibly for as long as a decade (Wohletz, 2000). In addition, between AD400 and 600 the Earth is believed to have experienced a sustained bombardment by meteor showers associated with the break-up of Comet Biela (Clube, 1992). This would also have caused a climatic downturn as the additional debris in the atmosphere would block out incoming sunlight, thereby reducing surface temperatures.

There are plenty or records of cold spells and snow for Ireland from this period, and some are associated with multiple fatalities, indicating the severity of the events. However, the event descriptions are generally very short and lacking key details such as dates and duration. There is also likely to be some duplication of events because of difficulties in clearly identifying the year in the original records (different monastic annals having slightly different dates for the same event) and these possible duplications are indicated below.

AD582 Great snow (GBCO, 1852).
AD587 A great snow (GBCO, 1852).
AD588 A huge snow and a great frost this year (Britton, 1937). Possibly the same event as AD587.
AD634 A great snow caused the death of many in Ulster (Britton, 1937).

AD638 A great snow killed many in the plain of Breagh in Co. Meath (GBCO, 1852).

AD669 A great snow and great scarcity (GBCO, 1852).

AD670 Snow killed many cattle (Britton, 1937). Possibly the same event as AD669.

AD695 The sea between Ireland and Scotland and the lakes and rivers of Ireland, frozen. The people of Ireland and Scotland paid reciprocal visits over the ice (GBCO, 1852).

AD748 An unusually great snow so that nearly all the herds were destroyed in Ireland and afterwards a noteworthy drought occurred in the world (Britton, 1937).

AD759 5 February: a great snow (GBCO, 1852).

AD761 Great snow (GBCO, 1852).

AD762 Great snow (GBCO, 1852). Possibly the same event as AD761.

AD763–764 There was a great snow with intense frost, not to be compared with any in former ages. It covered the earth from the beginning of winter almost until the middle of spring and through its rigors the trees and vegetables mostly withered away and many marine animals were found dead. This was undoubtedly one of the severest winters known to history (Britton, 1937).

AD766 Snow beyond measure (GBCO, 1852).

AD779 April: a great snow (GBCO, 1852).

AD788 3 May: great snow (GBCO, 1852). A late date for snow.

AD798 Great snow in which people and cattle perished (GBCO, 1852).

THE 'MEDIEVAL WARM PERIOD', AD800–1300

The temperatures returned to levels that were comparable to those of the mid-twentieth century during this period. There was a general retreat of glaciers, sea ice and ice sheets. There was less rain, leading to recurring droughts and water shortages. This milder weather facilitated the settlement of Iceland and Greenland by the Vikings and their other numerous voyages around Europe, and the growing of grapes in England. The Irish annals confirm this generally better climatic picture. They suggest that Ireland experienced heat waves, droughts and summer lightning. There are also many records of cold spells and severe snowfalls and the dating and detail of the events is a little better.

AD816   25 December–22 February 817: severe winter in Ireland. A marvellous frost and great snow persisted and the bogs were crossed dry-shod and many rivers and lakes in the same way. Stags were taken without being hunted. Much material and parts of houses were carried across Lake Erne out of Connacht (Britton, 1937).

AD821–2   Severe winter, notably in Ireland. Great frost so that the seas and lakes and rivers were frozen, that herds, horses, flocks and burdens were conveyed across them (Britton, 1937).

AD847   1 February: a great snow in Ireland (GBCO, 1852).

AD854   10 May: snow as high as men's girdles (GBCO, 1852). This is a very late date for snow in Ireland.

AD855   10 December–21 January 856: severe winter in Ireland and Scotland. Great ice and frost so that the principal lakes and principal rivers of Eirinn were passable to pedestrians and horse-riders (Britton, 1937). This is possibly the same event as AD854.

AD895   Great snow and a great mortality (Britton, 1937).

AD917   Severe winter in Ireland. Great frost in this year and prodigious snow which inflicted slaughter on cattle. Snow and very great cold and marvellous frost in this year, so that the more important lakes were crossed, and as a result of which in all Ireland there was a pestilence of herds, birds, horses, salmon and likewise a mortality among the flocks (Britton, 1937).

AD941   A great frost in Ireland so that the lakes and rivers were passed over and foreigners lay waste to the island of Mochta by crossing over the ice (Britton, 1937). The island of Mochta, which was originally in Co. Louth, is now in Co. Meath. This was a small island on the River Dee with the monastery of St Mochta on it, in addition this area has been drained at some time in the past so it is no longer a lake either (Tempest, 1944). One of the last two events may have been contributed to by the eruption of Langjökull in Iceland in AD925±25 (Simkin, 2002).

AD955   Great frost and snow, destructive to cattle (GBCO, 1852).

AD959   March: a great snow happened causing a great mortality upon cattle by both snow and distempers (Britton, 1937).

AD960   Snow and diseases (GBCO, 1852). Possibly the same event as AD959.

AD963   Great famine and cold and scarcity of corn (Britton, 1937). One or more of these events may have been exacerbated by the eruption of Ljósufjöll in Iceland in AD960±10 (Simkin, 2002).

AD1008   6 January–28 March: severe winter in Ireland with great frost and snow (Britton, 1937).

AD1011   11–14 February: a great snow in Ireland (Britton, 1937).

AD1022   There fell a great and wonderful snow (GBCO, 1852).

AD1026   Cold weather in Ireland and people crossed over the ice to the island of Mochta for the second time (Britton, 1937).

AD1029   1 February–17 March: great snow (GBCO, 1852).

AD1031   Heavy snows in Tirconnel (Co. Donegal and parts of Cos. Sligo, Leitrim, Tyrone, Fermanagh and Derry). The unusual expression 'the prey of the snow' is used, possibly indicating severe casualties (Britton, 1937).

AD1046   8 December–17 March 1047: severe winter with heavy snows. Snow fell and remained from the calends of January to the festival of St Patrick, whence it was called the great snow (Britton, 1937). There was also a destruction of men, cattle and of the wild animals of the sea and birds (GBCO, 1852).

AD1075–1076 Great frost and snow in the end of this and the beginning of the next year, (GBCO, 1852).

AD1078   Snow and great frost (GBCO, 1852).

AD1088   Great snow in this year (GBCO, 1852).

AD1092–1093 A huge snow fell and a great snow and frost this year so that the lakes of Ireland were congealed (Britton, 1937).

AD1095   3 January or 28 March: great snow fell on the Wednesday after the calends of January, which killed a multitude of men, cattle and birds (Britton, 1937). The GBCO (1852) records that there was great snow on the Wednesday after Easter, which destroyed people, birds and cattle, and there was a severe frost. Clearly, there is an error here in one of the accounts and this epitomises the difficulty of dealing with very old calendars.

AD1098   Great snow in this year (Britton, 1937).

AD1105   25 January: a great snow about 1105 (Britton, 1937).

AD1107   23 March: snow fell for a day and night which caused a great destruction of cattle in Eirinn (Britton, 1937).

AD1110–11 The snow of the birds. Great frost, so that the drovers passed dry-footed the lakes of Eirinn and a severe winter (Britton, 1937). Extreme ill weather, frost and snow in which both tame and wild beasts perished (GBCO, 1852).

AD1114 18 December–15 February 1115: severe winter. Very great sweeping snow and frost, so that the drovers were wont to pass dry-footed over the principal lakes of Eirinn and which killed a multitude of cattle, birds and men (Britton, 1937). Although similar to the description of AD1110–1111, this is definitely a separate event.

AD1156 A great snow and severe frost in winter this year, so that the lakes and rivers of Ireland were frozen over. Such was the intensity of the frost that Roderic O'Conor transported his ships and fleets across the ice from the margin of the shores of Galvia to Rinduin (this is a little puzzling, as Galvia is the old name for the Corrib and hence Galway, whereas Rinduin (Rindoon) is on Lough Ree on the Shannon (maybe what is meant is that both were crossed on the ice and the frozen land in between)). Likewise, numbers of large birds of prey were killed in Ireland by the intensity of the snow and frost (Britton, 1937). It is suggested by GBCO (1852) that most of the birds of Ireland perished, not just birds of prey.

AD1193 28 March and 4 April: great snow (Britton, 1937).

AD1200 A cold, foodless year, the equal of which no man witnessed in that age (Britton, 1937).

AD1205 1 January–17 March: great frost and snow (Britton, 1937).

AD1233 25 December–6 January 1234: a great frost in this year so that the lakes were bound up and people, horses and flocks went over many lakes. Great snow and frost afterwards, so that men and horses under burdens would pass over the principal lakes and rivers (Britton, 1937).

AD1245 6–25 December: poisonous snow fell on this night, which took off the heels and toes of those who walked in it: and this snow did not disappear until Christmas arrived (Britton, 1937). Clearly this is another case of acidic volcanic aerosols and particles mixing with the snow (if it was snow at all, as it could have been volcanic ash on its own). The well-known eruption of Katla in southern

Iceland in this year was the cause of this unusual snow (Simkin, 2002).

AD1247 Winter until 11 July: the winter was stormy, cold and wet, and continued so until 11 July, insomuch as the gardeners complained that winter was turned to summer and summer to winter, and that they were like to lose all and be undone (Tuckey, 1837).

AD1251 Great frost in the early winter so that the lakes and the bogs and the waters were frozen (Britton, 1937).

AD1278 25 December–February 1279: great snow and frost (GBCO, 1852).

AD1281 25 December–1 February 1282: very great snow (Britton, 1937). Possibly the same event as AD1278–9.

THE CLIMATIC DOWNTURN, AD1300–1450

After around 1300, the climate got worse, with a gradual drift back to colder and wetter conditions. The glaciers and ice-sheets re-advanced, and there was a southward movement of sea-ice. The Viking colonies in both Greenland and Iceland started to struggle and, in England, wine production started to decline and conditions for agriculture all across Europe deteriorated. In Ireland, the annals track the deterioration through references to famines caused by lower temperatures and particularly disastrously high rainfall levels. Two particularly horrific periods (both associated with prolonged rainfall) stand out, 1315–17 and 1348–9, when climate-derived famines most likely caused the deaths of several hundred thousand people, although there is no way of being sure about any fatality figures for this period.

AD1318 Snow, the like of which was not observed for a long time, fell this year (Britton, 1937).

AD1335 There was such great snow in the spring of this year that the most of the small birds of Ireland died (GBCO, 1852).

AD1336 A great plague of snow and frost (GBCO, 1852).

AD1338 30 November–22 January 1339: this year was exceedingly stormy and harmful to men and animals because there was snow and ice. Agriculture ceased on account of the snow and ice which continued abundantly until that time. Also, in parts of Ireland the frost was so severe that the river of

Dublin, the Liffey, was frozen so that many people disported themselves upon the ice, indulging in games and races: then upon the ice a fire of wood and grass was kindled, the ice accompanying this same frost was of wonderful depth (Britton, 1937). This is very similar to the frost fairs we associate with the Christmas stories of Charles Dickens, which date from the mid-nineteenth century.

AD1433 6 November–13 January 1434: great frost in this year and numerous herds of cattle and horses and people and packhorses used to go upon the chief lakes of Ireland during the frost and so on. The Fir-Manach (from Co. Fermanagh, but possibly originally from Leinster) sent their movables over Lough Erne westwards and they had no vessels but the solid ice on the lough (Britton, 1937).

THE 'LITTLE ICE AGE', AD1450–1850

As its name suggests, this was a period of lower temperatures and harsher climatic conditions, with much snow, cold spells and ice, and significant glacier re-advances. The severest part of the Little Ice Age was from 1670 to 1710, when temperatures fell to averages several degrees below current values. This cooling coincided with almost no sunspots, and is known as the Maunder Minimum. Low sunspot numbers are associated with reduced solar output. The Little Ice Age was episodic in nature, with cold spells alternating with phases of milder weather, but always against a backdrop of generally low temperatures and increased rainfall. There were massive increases in the extent and range of sea ice. The collapse of the Viking colony in Greenland and the near collapse on a number of occasions of similar settlements in Iceland are a result of the severe sea ice. By the early eighteenth century, the sea ice drifted so far south that on several occasions between 1690 and 1728, Eskimos, and in one instance a polar bear, made it all the way to Scotland (Dawson, 2009). Rivers and coastal inlets across north-western Europe froze regularly, more than once a decade in some cases. During the winter of 1683–4, slabs of ice three miles across appeared in the English Channel, and the Thames froze at least twenty times during these years. Even the Mediterranean Sea around Marseille froze in 1595 and 1638, indicating that the big chill was not just a north European phenomenon (Hickey, 2008).

In Ireland, there is plenty of evidence of the climatic severity of the latter part of this period, although we are reliant now on the census data for most of the early records at least until the eighteenth century, as Britton's (1937) work only goes up to AD1450. As a result of the struggles and wars of the sixteenth and seventeenth centuries and resulting famines and population declines, there are only a few records from the 1400s, 1500s and 1600s. From AD1780 to 1850 there are lots of records of more mundane snowfalls and brief cold spells, so that the chronology from then on is much more populated with events. (All dates below are 'AD'.)

1462   A great frost that slaughtered many flocks of birds (GBCO, 1852).

1465   An exceeding great frost and snow (GBCO, 1852).

1517   Winter: great frosts, and horses crossed the rivers of Ireland, the River Lee froze for weeks at a time (GBCO, 1852). In the winter of this year, there happened a great frost, so that all the rivers of this county (Cork) were frozen up for several weeks, particularly the Lee and Avenmore, i.e. the Blackwater (Smith, 1839).

1541   Great inclemency of weather, both frost and snow (GBCO, 1852).

1600–1 A hard frost (GBCO, 1852).

1635   January–February: a great store of snow did fall to the great damage of the cattle, chiefly in the northern parts (where it did snow most exceedingly) so as the people were put to hard shifts to bring their cattle in safety to their folds and other covered places. It lingered long on the mountains, including the Wicklow Mountains, for many days and weeks after it had thawed and quite vanished on lower ground (Boate, 1652). It is interesting to note that Boate also commented that snowfalls were not that common in Ireland and the effects of the Little Ice Age may not have been as severe as in more continental climates. This relative amelioration is a product of our maritime climate, which tends to reduce the occurrence of extremes.

1641   Most bitter cold and frost (GBCO, 1852).

1683   A most severe frost, the River Lee froze for weeks at a time (GBCO, 1852). Carriages passed over the ice from the ferry slip to the east marsh in Cork city (Smith, 1839).

1691   January and February: a great frost (GBCO, 1852).

1692   19 January–February: a great frost began in Ireland, and continued until the middle of February (Tuckey, 1837). Most likely the same event as that of 1691.

1693   Some days of hard frost (GBCO, 1852).

1708   December–March 1709: very severe frost throughout Europe, but scarcely felt in Ireland but Dublin had a harder winter than usual (Lowe, 1870). The problem with Lowe is that he is the least reliable of the chroniclers, to the point where some events were clear fabrications or duplications of earlier events, with no supporting evidence and which do not appear in any other chronicle. One of Britton's main tasks was to carefully cross-check the work of Lowe and hence our low opinion of his reliability as he records events (Britton, 1937). Britton's work only covers the period up to AD1450 and it is likely that Lowe got more reliable later, as the events in his chronicle were more recent and verifiable.

1715   March: frequent frosts (GBCO, 1852). There was a great fall of snow which lay for two months according to Tuckey (1837).

1719   A frosty winter (GBCO, 1852).

1720   A wet, cold winter with frost and sleet (GBCO, 1852).

1726   January: a month of the hardest weather for snow (GBCO, 1852).

1728   23 June: ice was found on the River Liffey (GBCO, 1852). This gives some idea of just how bad some of the summers were, not to mention the winters. This is also possibly down to the influence of the Öræfajökull eruption of 1727 (Simkins, 2002), which might help explain the occurrence of ice in June in Dublin, which is extremely unusual, even allowing for the fact that this event occurs in the Little Ice Age.

1728   December: frost and snow was more severe and of longer duration than for many years past (GBCO, 1852).

1739   26 December–September 1741: this event was produced by a combination of the Little Ice Age and a major volcanic eruption, most likely in Siberia, which caused an enhanced temperature drop. Dickson (1997) examines this event in detail and with a startling conclusion as regards the mortality levels.

1740–1850

This period was one of extremely disturbed weather, with the direst of consequences. It started with a period of intense cold that lasted from late December 1739 to May 1740, with little precipitation. The intense cold froze rivers and lakes (including the Liffey, Lagan, Shannon and Boyne and Lough Neagh) for periods of weeks at a time, and caused the deaths of numerous domestic and wild animals and the virtual drying up of streams and rivers. The temperatures were so low that liquids froze indoors. From 26 December 1739 to 4 January 1740, the River Lee was frozen by one of the hardest frosts in memory. This is usually called 'the hard frost', after which a great scarcity followed. During the time it lasted, tents were fixed on the river from the North Strand to Blackrock, near Cork, and several amusements were carried on there, which continued even after the commencement of the thaw (Tuckey, 1837).

The single outdoor temperature reading taken during these seven weeks of aggressively cold weather was 32°F or −36°C, which is some 17°C lower than the modern instrumental record for Ireland. Even allowing for some inaccuracy, this figure gives an idea of just how cold it got during this period. Manley's *Central England Temperature Record* (1974) shows 1740 to have been the coldest year between 1659 and the present, with a mean annual temperature of just 6.8°C. This is the longest instrumental temperature record in the world. To put this figure in context, the lowest twentieth-century mean annual temperature was 8.5°C, which occurred in 1963, whereas the 2009 mean was 10.1°C, nearly 3.5°C higher (Figure 9.1).

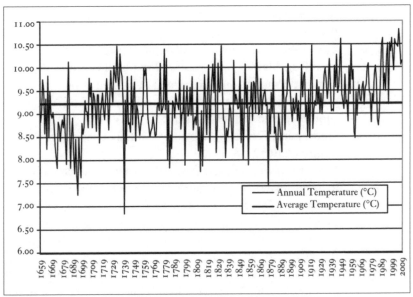

Figure 9.1 Manley's Central England Temperature Record (°C) from 1659 to 2009 (after Manley 1975 and updates).

Deluge

The harvest that year was very poor because of the late planting and the lack of precipitation. By late 1740, food prices had risen so much that the first deaths from starvation were recorded. The winter of 1740–1 was particularly harsh, with frequent snowfalls and frosts and storms. The death-toll escalated rapidly throughout the first half of 1741. Relief came via a better harvest in 1741 and the arrival of five grain ships from America.

This Siberian winter was followed by a prolonged drought, famine, typhus, dysentery and, in towns, the threat of civil unrest. Between 310,000 and 480,000 people died during the crisis, out of a population of 2.4 million, a mortality-rate greater than that experienced during the Great Famine of 1845–51. Writing in April 1741, as disease savaged the malnourished population, the Revd Philip Skelton from Co. Monaghan stated that the dead had been eaten in the fields by dogs for want of people to bury them. Thousands in the barony perished, some of hunger and others of disorders occasioned by unnatural, unwholesome and putrid diets, such was the nature of the calamity (Dickson, 1997).

1744   The northern part of Ireland was covered with snow for weeks (GBCO, 1852).

1745   A great fall of snow that smothered numbers of cattle and sheep (GBCO, 1852).

1749   June: a hoar frost (GBCO, 1852). A very late date, even for this type of frost.

1750   23 October: a memorable fall of snow (GBCO, 1852).

1752   23 and 25 March: snow recorded in Dublin (Lowe, 1870).

1752   9 and 10 October: snow recorded in Dublin (Lowe, 1870).

1753   3 and 19 January, 5–7 and 11–12 February and 1 March: snow recorded in Dublin (Lowe, 1870).

1758   January and February: frosty, a great fall of snow in the latter month (GBCO, 1852).

1766   A frosty spring. The greatest fall of snow ever remembered, in some places it was more than 4.6m deep, numbers of sheep and several travellers were lost (GBCO, 1852). In February, the rivers, loughs and canals being frozen, there was much skating until 8 February, when the frost began to thaw (Tuckey, 1837).

1767   2–17 January: intense but short frost, attended by a great fall of snow (GBCO, 1852). On 2 January, there was much skating on the lough in Cork. By the 12th the frost had continued with such severity, that the poor tradesmen and manufacturers were entirely idle, not being able to follow their occupations. For want of something better to do, many resorted to the lough to amuse themselves by skating. On the 13th, however, the frost began to thaw. For the time it continued, it was supposed to have been as severe as that in 1739, but no material damage was caused to the country. On 17 January, the frost set in again with great severity. The roads were almost impassable. The snow was seven or eight feet deep (Tuckey, 1837).

1768　11 January: severe frost, people walked on the Liffey (GBCO, 1852). There had been for some days the greatest fall of snow for forty years. In some places, it was nearly 2m high. The horse of a gentleman, who was riding from Bandon, Co. Cork, sunk in it so deep that three or four men were employed to dig him out with spades and shovels (Tuckey, 1837).

1771　3 January–18 April: almost continued frost with snow and an easterly wind (GBCO, 1852).

1776　31 January: the Dublin post, which should have arrived the 29th in Cork, did not come till one o'clock this day, on account of the great fall of snow which rendered the roads almost impassable (Tuckey, 1837).

1779　3 May: there were several showers of hail, succeeded by a heavy fall of snow, the weather had been remarkably cold and severe for some time before this date (Tuckey, 1837). This was another one of these very late events not untypical of the Little Ice Age.

1783–4 The Liffey was covered in solid ice (GBCO, 1852). 7 February 1784: a heavy snow was congealed by a severe frost. The lakes and rivers were frozen over. Excessive cold with snow from 4.6 to 6.5m deep (GBCO, 1852). There was the severest frost since the year 1739 (Tuckey, 1837). This event relates to the famous Laki fissure eruption, which lasted over eight months between 1783 and 1784 and the Eldgjá fissure eruption (Simkin, 2002; and see below, ch. 11).

1786　The frost uncommonly severe (GBCO, 1852).

1788　14 December–17 January 1789: on the night of 14 December and the following day, there was a great fall of snow in Cork city, which was succeeded by so severe a frost that the south channel from Parliament Bridge to the edge of Lapp's Island was frozen, and the navigation of ships greatly impeded. The frost having continued for the next four days, several persons were hardy enough to skate this day on the river in the south channel of the Lee (Tuckey, 1837). A severe frost set in on 29 December (GBCO, 1852). On 17 January, however, a combination of a thaw and heavy rain produced a spectacular flood in Cork city, one of the biggest on record (see above, ch. 5). Great frost and snow in Dublin and the south of Ireland in 1789 (GBCO, 1852).

1798 Great snow (GBCO, 1852).

1799 January: a severe frost was succeeded at the end of January by a heavy fall of snow (GBCO, 1852).

1799 April: a heavy fall of snow from Thursday night to Saturday morning (GBCO, 1852).

1801–2 Intense frosts (GBCO, 1852).

1802: May: severe frosts (GBCO, 1852). This is another example of a very late occurrence of frost.

1803 12 January–February: on 12 January an intense frost set in and at the end of February there was a considerable fall of snow (GBCO, 1852).

1807 January: great snow, the roads to the north and east impassable (GBCO, 1852).

1807 19–20 November: the snow ten feet deep in the neighbourhood of Dublin, winter intensely cold (GBCO, 1852). A disastrous blizzard swept the country and many people were killed. Two transport ships were wrecked on the east coast. Heavy snow prevented the crews from realising how close to land they were (Rohan, 1986).

1809 January: travel between Dublin and the country arrested by a great fall of snow early in January (GBCO, 1852).

1812 March: much snow (GBCO, 1852).

1813 25 December–31 March 1814: a severe frost accompanied with much snow. The snow was 1.5m deep in Dublin and much deeper in the country, the old Londonderry Bridge was carried away by the ice (GBCO, 1852). It was noted by Tuckey (1837) that most serious inconvenience resulted from the suspension of travelling, occasioned by a great fall of snow, which far exceeded any similar calamity, within the memory of the oldest inhabitant of this county of Cork.

1815   November–10 December: in the end of November, the frost set in unusually steady. The weather was very severe and, there having been a gradual thaw the previous day, 9 December, there was a considerable flood in the river (Tuckey, 1837).

1816   This year was known as 'the year with no summer' and was associated in Ireland with an unseasonably cold summer. This was caused by the eruption of Tambora in Indonesia in 1815, which caused a global downturn in temperature (see below, ch. 11). The worst effects seem to have come in the summer and not the winter.

1819   October–November: great snow in Dublin in October, the same not observed so early for eighteen years previously. An early frost occurred in November (GBCO, 1852).

1820   4–22 January: this season was remarkable for the severity of the weather, it being generally supposed that the frost equalled that of the year 1739, many of the largest rivers in the country having continued frozen for some days, it became necessary to have recourse to sledges to break the ice, to enable the farmers to procure water for their cattle; the River Lee was crossed in various places by foot passengers. The cold weather continued for five weeks, in consequence of which the poor were reduced to the greatest poverty and misery, many having been compelled to pawn the wretched coverings of their beds, to obtain the means of subsistence for their starving families. The weather increased in severity, the snow was 1m deep in the streets of Cork and in various parts of the country, the roads were scarcely discernible. On 21 January, a poor woman perished in a heap of snow near Gallows-green. From 22 to 24 January, a thaw combined with continuous rainfall produced another flood in the city of Cork (Tuckey, 1837).

1821   A severe frost early in the year. The navigation of the canals suspended, and intercourse with the metropolis both by land and water interrupted. Towards the end of May, the thermometer often fell to zero (GBCO, 1852).

1823   January: a most severe month, immense quantities of snow fell. Also a great frost occurred. As a result, the mails were unable to travel and the navigation of canals was interrupted (GBCO, 1852).

1825   October: great snow in Dublin (GBCO, 1852). This is a relatively early date for snow in Ireland.

1826   The winter set in early and was marked with unusual severity. There was frost on 8 November (GBCO, 1852).

1827   January: a heavy fall of snow followed by frost (GBCO, 1852).

1829   Cold year. A cold spring with north and northwest winds, summer also cold, winter inclement beyond the average. Frost occurred in December (GBCO, 1852).

1830   June was described as being as cold and nipping as March (GBCO, 1852).

1830   December: there was much snow and severe frost. Some roads blocked up for weeks (GBCO, 1852).

1833   March: much snow on the Dublin and Wicklow Mountains (GBCO, 1852).

1833   15 October: much snow on the Dublin and Wicklow Mountains (GBCO, 1852).

1834   January–April: severe weather in Ireland continued till after the first week of April (GBCO, 1852).

1835   The spring was very severe (GBCO, 1852).

1836   Weather bleak, raw and cold. On 7 November, several hills were covered with snow (GBCO, 1852).

1837   January: great snow (GBCO, 1852).

1838   February: snow storms on 16 and 18 February, 20cm of snow fell at a time. A great storm occurred at Cork in the latter end of February and was attended with a heavy fall of snow (GBCO, 1852).

1840   Much snow (GBCO, 1852).

1841   January–February: a heavy fall of snow, the canals had frozen some 4cm thick and the Liffey froze. It continued with slight thaws to the end of the month and was followed by a deep fall of snow on 2 and 3 February, with parts of Cork covered to a depth of 1.1 to 1.6m (GBCO, 1852).

1842   Great cold in Limerick. Much snow also (GBCO, 1852).

1843   January: snow (GBCO, 1852).

1843   April: a heavy fall of snow in Dublin (GBCO, 1852).

1844   April and May: heavy showers of snow and hail did much injury to fruit blossoms and to vegetable gardens on 18 April. Frosts occurred almost every night in May and were reported to have injured the flax and potatoes (GBCO, 1852).

1845   28 January: a great snow storm at Ballinrobe, Westport and Tuam (GBCO, 1852).

1845   14 March: a severe frost, the River Liffey frozen over above King's Bridge followed by a heavy snow storm on the 19th and showers of sleet in the midst of the dog-days (i.e. somewhere between 24 July and 24 August) at Limerick (GBCO, 1852).

1845   8 April: snow on the Dublin Mountains (GBCO, 1852).

1845   August: frosts of unusual character during some nights (GBCO, 1852).

1846   3 March: a great snow storm in Dublin and at Killarney (GBCO, 1852).

1846   8 April: frost and snow in Belfast (GBCO, 1852).

1847   8 February: a heavy fall of snow in Tipperary and Belfast. This was followed by severe frost on the night of 12 February (GBCO, 1852).

1847   30 March: snow (GBCO, 1852).

1848   May: a sharp frost the first week of May (GBCO, 1852).

THE GLOBAL WARMING PERIOD, AD1850-PRESENT

There are a number of distinct phases that can be identified from AD1850 onwards. Between 1850 and 1920, temperatures varied up and down and this could be seen as the transition from the Little Ice Age to global warming. From 1920, temperatures rose steadily up to *c*.1945, which remains the hottest year on record for Ireland. Between 1950 and the end of the 1970s or early 1980s, temperatures fell again, leading to an initial forecast by many climatologists of a return to ice-age conditions. Once this concern began to be investigated, however, a contradictory picture emerged throughout the rest of the 1980s and, since then, temperatures have risen and risen, and seemingly at a faster rate. New record global temperatures were recorded on an almost yearly basis, with the 2000s being the hottest decade since instrumental records began in 1860. Despite the generally rising temperatures over this period, a number of significant cold spells with lots of snow and ice have occurred.

1850   7 May: the Dublin Mountains were white with snow (GBCO, 1852). This is the end of the cold entries from this source and it also marks the threshold into the global warming phase. This also explains why there are many more entries towards the end of this part of the chronology as there are numerous documentary sources, including newspapers, not to mention personal experience, of these later events.

1853   14 February: a violent snowstorm caused the ship the *Queen Victoria* to strike Howth Head, with the loss of 55 lives (Rohan, 1986).

1879   The intense and prolonged cold of the present winter was greater than any in the preceding ten years. The excess mortality was 20 per cent in Dublin and 12 per cent over the whole country, when compared to the average of the previous decade, notably those under five years and from fifty years upwards. The especially high mortality rates in Dublin were because of the percentage of the population under five or over fifty. No actual figures are given, but the implication is that there were thousands of fatalities nationwide. The Reykjanes eruption in Iceland may have contributed to the severity (Simkin, 2002).

1881   16 January: Markree Castle, Co. Sligo, recorded a temperature of −19.1°C, the lowest on record (bearing in mind that most meteorological stations were set up from the 1880s onwards; Rohan, 1986).

1886   28 February–2 March: a great blizzard with snow to a depth of 60cm struck northern Ireland and traffic was severely disrupted as a result (Rohan, 1986).

1891–2 Winter: noteworthy fall of snow (Rohan, 1986).

1894–5 Winter: noteworthy fall of snow (Rohan, 1986). As the record 1895 winter continued its wrath, Dublin was hard hit by heavy falls of snow on 8 January. On 13 January, an enormous snowfall hit Ireland. Train and tram services were cancelled. Those who ventured out were at great risk of becoming stranded (Bedford, 2008). In Killarney on 7 February, snow fell heavily for fifteen hours and on 10 February, Co. Kerry was hit by huge snowdrifts which blocked many roads. The ground was so hard that graves could not be dug, and the dead remained unburied until a thaw set in (Bedford, 2008). On 7 February at Mostrim, Co. Longford, a temperature of −17.8°C was recorded (Rohan, 1986).

1901   28 and 29 March: heavy snow in northern Ireland (Eden, 2008).

1902   8 and 9 February: widespread snow. The first three weeks of February were very cold and wintry, with frequent snow. The heaviest falls came on the 8th and 9th, leaving 30–35cm on the ground in parts of northern Ireland (Eden, 2008).

1904   20–23 November: outbreaks of snow, heaviest in Tyrone and Derry, with transport severely disrupted (Eden, 2008).

1906   25–30 December: two major snowfalls occurred. The first, accompanied by moderate winds, affected northern Ireland on the 25th–26th. Western Ireland reported 10–15 inches (37cms), a rare event according to Bedford (2008). On 27–28 December, the weather also brought 35–45cm of snow to northern Ireland (Eden, 2008).

1909–10   Cos. Kerry, Clare and Cork were covered with snow for several days (Rohan, 1986). From 14 to 19 January there was heavy snowfall with much drifting, chiefly on the 15th–16th in northern Ireland, resulting in blocked roads and railways. Rapid thaw from the 17th caused widespread floods, said to be the worst in northern Ireland for over twenty-five years (Eden, 2008).

1917   January–14 April: long cold and snowy winter. Unprecedented snowstorms hit Ireland during the first few days of April (Eden, 2008). There was widespread disruption throughout the country (McWilliams, 2009). The snows were described as the most severe of the present century. On 25 and 26 January, the south and west of Ireland experienced a severe snow storm which paralysed these areas. Snow to a depth of 60cm and drifts of up to 4.5m were recorded in Co. Mayo. In Carrick-on-Suir, the snows of 26, 27 and 28 January had rainfall equivalents of 25mm, 22mm and 10mm respectively. The low temperatures carried on into the following week and many roads were still impassable because the snow did not melt for days. An even more severe snowstorm occurred on 1 April, and many places were effectively isolated by this event, with snow depths of 1.3m and drifts of 3m (Rohan, 1986).

1933   The winter of 1933 was generally mild, except for a short cold spell in February and one in March (Bedford, 2008). On 23 and 24 February, snow made roads impassable for up to three days and some villages were cut-off for up to a week. The heaviest falls occurred in the midlands and the south.

This snow event was caused by a small depression of polar origin. On 25 February, this depression stalled over Northern Ireland, then deepened and then, as a result the precipitation which began as heavy rain, it turned to snow as it came up against a cold easterly wind sweeping in from Europe (Bedford, 2008).

1937   11–13 March: heavy snow. Worst hit was Northern Ireland, where all roads were closed for a time, and snow lay 25–35cm deep, with drifts of up to 3m (Eden, 2008).

1947   January–March (Table 8.1): this was one of the severest winters in living memory, with continuous snowfall isolating large parts of the country. It was known as 'the Big Snow' (McGreevy, 2010d). This event was caused by high pressure from Greenland extending to Scandinavia and causing cold easterly winds to become persistent over Ireland. When milder air from the southwest tried to get to its normal position over Ireland, the meeting of the two produced extensive snowfall. This event is notable for causing an increase in the death-rate among the elderly. The hardship associated with this event was exacerbated by the post-WWII rationing that was still going on, and this affected electricity supply, fuel and transport. Conditions were getting so bad that the army was putting in place plans to have field kitchens up and running in some of the poorer parts of Dublin, and it was suggested that reserves of coal in army barracks should be made available. The temperature did not rise above 5°C at Dublin Airport between 22 January and 7 March, with some light snowfalls occurring as late as 24 March. There were between twenty and thirty days of snowfall in most places, with notable snowfalls on 2, 8, 21 and 25 February and 4 March. The 25 February snowfall was described as 'the Great Blizzard' (Booth, 2010). Co. Wicklow received so much snow during this period that many people were trapped in their homes for days and even weeks.

1958   19–24 January: widespread snow, heavy over large areas, caused major disruption to traffic. Notable snow depths included 69cm at Aldergrove Airport in Co. Antrim. Drifts of 4–5m were widely reported (Eden, 2008).

1962   December–February 1963 (Table 8.1): this winter was very severe due to high pressure over the North Sea for two months, pushing a cold easterly airflow over Ireland, accompanied by repeated bands of snow. January was the coldest month on record for many stations, with a low of −16°C being recorded. The cold spell began in December 1962 and lasted for most of January and February, bringing heavy falls of snow to the east coast (McGreevy, 2010d). Between the 25th and 31st of December, but especially on the 30th and 31st, snow fell over most of the country. On the morning of the 31st, 45cm of snow was recorded at Casement Aerodrome. There was more snow on 15 January, and it was still on the ground in places until after 5 February (Rohan, 1986). The River Shannon could be crossed on foot near Limerick city for the first time in living memory.

1967   8 and 9 December: two polar lows brought two heavy snowfalls within thirty-six hours to Northern Ireland. Some 25–30cm of snow lay in Northern Ireland (Eden, 2008).

1969   9 February: 25–30cm of snow fell in Ulster (Eden, 2008).

1970   4 March: an unexpected snowstorm affected Northern Ireland (Eden, 2008).

1978   11–12 January: snow accumulated to a depth of 20–25cm in Northern Ireland (Eden, 2008).

1978   December–May 1979 (Table 8.1): this was one of the most prolonged cold spells to have affected Ireland since instrumental records began. Snowfalls occurred regularly and persisted, with hail showers right up to May. Mean temperatures between January and May were below normal, particularly in January, March and May. This was caused by west to northwest winds, which produced snow and sleet instead of the usual rain. New record low temperatures were set on 1–2 January at a number of meteorological stations, including Mullingar (−14.9°C), Kilkenny (−14.1°C) and Casement Aerodrome (−12.4°C; on 8 January 2010 that figure was equalled). On 2 January at Lullymore in Co. Kildare, a temperature of −18.8°C was recorded, the lowest of the twentieth century. This initial cold phase ended on 6 January, but was followed by other less severe phases (Rohan, 1986). There were significant falls of snow between 28 and 31 December 1978, with further snowfalls later in January. The frequency of snowfall can be seen by the fact that Dublin Airport recorded forty-two days of snow, with sixteen days of snow lying, compared to the normal values of sixteen and four respectively. Significant snowfalls occurred in Dublin on 14 February, causing massive traffic chaos and in many parts of the country on 16 and 18 March (Met Éireann, 2009).

1981　January: even the heavy falls of snow only lasted a week and temperatures were mild either side of it (McGreevy, 2010d).

1982　8–15 January (Table 8.1): there was widespread snow on 8 January, which was heaviest in eastern areas. In addition, there was significant drifting as a result of strong easterly winds. A severe cold spell occurred after the snowfall and lasted until 15 January, when the snow finally started to melt (Rohan, 1986).

1984　18–27 January: heavy snow fell repeatedly over Northern Ireland (Eden, 2008).

1997　4 May: the Sunday of the Bank Holiday weekend was warm (up to 25°C), but later in the day a depression streamed south from the Arctic Circle, bringing sleet and snow to many areas. Afternoon temperatures dropped to 8°C. It had not snowed this late in the year in Ireland since 1979, but on the morning of the 4th, snow lay in Cos. Sligo, Kerry, Mayo, Galway, Leitrim and Donegal. It was up to 4cm in places. The season's new lambs were about three weeks old, and while they were no longer considered to be at risk, this abrupt change must have come as a shock (Bedford, 2008).

1998　5 January: 25mm of heavy rain in Munster turned to snow as a depression made landfall in western Ireland. Four hours later, there was a 14cm blanket of fresh snow across Munster, but observers noted that neighbouring hills in Limerick and Tipperary were perfectly clear (Bedford, 2008).

1998　11–15 April: appreciable depths of snow were reported in Northern Ireland (Eden, 2008).

2000　27 and 28 December: heavy snow affected Northern Ireland, in Belfast snow lay 20cm deep, and falls of 25–35cm were reported in parts Co. Antrim (Eden, 2008).

2001　January: cold weather.

2001　4–7 February: heavy snow fell across much of Northern Ireland (Eden, 2008).

2004　26 February: the month ended with a week of snowy conditions. An innocuous-looking depression crossed over

England and stalled over the North Sea, drawing Arctic air from Scandinavia. By Thursday, the snow had moved into Northern Ireland, closing many roads (Bedford, 2008).

2006   1 March: 13cm of snow fell over parts of Northern Ireland, with road and school closures (Bedford, 2008).

2009   January–February: temperatures in January were the lowest since 2001 in most places and the lowest in Cork city since 1997. This spell of cold weather broke the sequence of eight previous mild winters going back to 2001. The impact of this cold weather was dramatic, and there was a 10 per cent rise in the death-rate among the over-65s during the first quarter of this year. This is an extra 613 deaths when compared to the three previous years. This shows the dramatic impact of cold weather on the elderly. The partial freezing of Wexford Harbour occurred for the first time in over twenty-five years in January. In February, heavy snow in and around the Wicklow Mountains made many roads impassable.

CONCLUSIONS

The record of cold spells, snow and ice, although covering a very long period, must be considered very patchy at best. The records of these events have to be read and interpreted with great care, particularly in assigning causation and inferring a major climate signal. It is suggested that, based on the historical evidence at present for the 'Little Ice Age' in Ireland, a prolonged cold spell did not take place and that it is much more sporadic when it occurs. Snow, like rain, exhibits widespread spatial variability, and in terms of occurrence, so that even a change of 0.1 °C can decide whether precipitation is going to fall as snow or as rain. These spells continue to the present day, as the 2009–10 winter spell has shown. This will continue to be the case into the future, but as global warming continues, the frequency and severity of these events will gradually diminish for Ireland.

CHAPTER 10

Burst pipes and flooding again

INTRODUCTION

The impact of the cold spell and its aftermath were considerable. During the cold spell, there were many accidents involving pedestrians and vehicle drivers and passengers, as well as significant transport disruption. The aftermath, when the thaw set in, was different in character. In particular, issues to do with water were at the forefront, including further flooding, burst pipes leading to considerable damage, particularly to ordinary households, and, surprisingly, water shortages. This chapter deals with the impact of the cold spell and the aftermath. It would be impossible to list all that occurred, so examples from particular days are used to illustrate the type of impacts that were common.

THE INJURIES

Although there were no confirmed fatalities from the cold spell, there is no doubt that the mortality rate among the elderly in particular increased, and this is a common feature of all major cold spells. A 10 per cent increase in the mortality rate would not be unrealistic. There were numerous injuries as a result of falls and car accidents caused by the treacherous road conditions.

The cold spell across Europe was more devastating than in Ireland. By 8 January, it was estimated that 122 people had died as a result of the cold in Poland, 22 in Britain and 9 in Germany. Many of these fatalities were homeless people, and a further 22 individuals were killed by avalanches in the Swiss Alps (Beesley, 2010).

One of the major impacts of the cold spell was the huge numbers of people who were hurt by falls on icy pavements and roads. Many suffered broken bones and other serious injuries. The Health Service Executive (HSE) reported a 65 per cent increase in fractures, putting accident and emergency departments under intense pressure. On 6 January, for example, as a result of a heavy snowfall and the ongoing icy conditions, there were so many falls that there was a 2–3 hour waiting time for ambulances in Dublin and elsewhere.

One farcical and yet tragic outcome of the cold spell relates to clearing pavements of dangerous ice and snow and the legal implications of this. There are potential liabilities for people who clear ice and snow from in front of their properties. This has come about because of a legal grey area. Currently, if someone were to fall on the cleared area, the person who cleared it is potentially liable for damages if that injured person were to sue. This is a ridiculous situation, and in other countries this obligation to clear pathways and pavement without liability is well-established. How many people could have been saved from falls which caused broken bones and other serious injuries had this legislation been in force? The Department of Transport is considering imposing a statutory obligation on people to clear snow and ice from footpaths outside their premises, without incurring any liability for negligence (Minahan, 2010).

THE CO. CLARE RESCUE

One man had a very lucky escape in one of the more dramatic rescues of the cold spell. A 78-year-old farmer got stuck in deep mud in a field near Quin, Co. Clare, while out feeding cattle on 31 December. A search began for the farmer when he failed to return by 11pm that day. The coast guard helicopter from Shannon was called in to assist with the search, and the farmer was eventually spotted, primarily because he was wearing a high-visibility jacket, which probably saved his life. The coast guard stated that the man was so embedded in the mud that they had to use a winch to get him out. The man was already suffering from hypothermia due to the mud, fifty-knot north-westerly winds and driving rain. He was brought to Limerick Regional Hospital, where he made a full recovery from his ordeal (Siggins, 2010).

ROAD CONDITIONS

Throughout the cold spell, road conditions were treacherous, especially on untreated roads in every part of the country, whether in rural areas or in housing estates. To outline the road condition problems would take a whole book in itself, so instead a snapshot of the conditions on a number of particular days during the cold spell is presented below. These snapshots are typical of what occurred on a daily basis. Right at the start of the deterioration into the cold spell on 22 December, road conditions were treacherous in most of the country, especially in Cos. Donegal, Sligo, Roscommon, Leitrim, Cavan, Monaghan, Longford, Westmeath and Clare.

On New Year's Day, the M50 and N11 were very dangerous. Driving conditions were desperate in many parts of the country due to the snow and ice, but especially in east Leinster. Abandoned crashed vehicles on Tinkers Hill, the Strawberry Beds and Knockmaroon Hill near Lucan, Co. Dublin, caused these minor roads to be closed. Islandbridge, Chapelizod and Parkgate Street Gates in the Phoenix Park were also closed. Roads were particularly bad in Laois, Offaly, Longford, Roscommon, Galway, Meath, Louth, Kildare, Dublin and Wexford, as well as the M4 between Mullingar and Kinnegad, the M1 between Carlingford and Ardee and the junction from Dublin Airport onto the Swords road. Conditions in Cork were so bad, with temperatures of −3°C, that Gardaí were urging motorists not to travel. The N8 at Watergrasshill was particularly treacherous, while Kanturk had freezing fog. The old N6 at Loughrea was icy, with black ice in many parts. Freezing conditions in Kerry caused the closure of the Connor Pass between Dingle and Tralee, also along the Kerry/Limerick border, where conditions were desperate on the roads. Upland areas suffered fog all day.

On 12 January, the snowfalls in many parts of the country caused numerous crashes and tailbacks, particularly in Cos. Wexford, Wicklow, Dublin, Carlow, Kilkenny, Sligo and Mayo. Wicklow Mountain Rescue had to rescue a Sky News team stuck on a remote mountain pass near Aughavannagh, presumably much to their embarrassment. Ice caused massive road damage, including deep potholes on several routes in Cork, especially the N28 Cork to Carrigaline road and between Tallow and Middleton, as well as in Co. Meath between Kilmoon Cross and Duleek, and Ratoath and Dunshaughlin. The Vico Road in Dalkey closed due to a collapsed wall.

On 14 January, Galway had its worst driving conditions since the start of the cold spell as a result of freezing fog and icy roads and a rapid temperature drop between 4am and 6am. Gritters only started to deal with the ice at 7am. This resulted in 2–3-hour tailbacks some 10km long on approaches to the city. Conditions in Limerick city and county and in Co. Clare were no better. A minibus overturned near Horse and Jockey in Co. Tipperary, causing non-serious injuries to passengers including children. The N8 in Co. Tipperary was also very dangerous due to ice. The impact across Europe was very serious indeed, hampering transport of all kinds and causing numerous traffic accidents and breakdowns.

There was great concern among local authorities about getting extra funds to cover the cost of repairs to roads, given the scale of the damage from both the flooding and the cold spell. Due to the current economic conditions, however, the government advised local authorities to set aside contingency funds for weather-related events. Subsidence on a major scale was reported by the AA on roads in parts of Cos. Cavan, Cork, Kilkenny, Meath, Monaghan, Offaly, Sligo, Tipperary and Wexford, indicating the scale of the road damage that had occurred. To put this in perspective, Cork County Council alone had to put in place funds of €6m just to meet the emergency repairs needed for roads and water supply networks. Repairing damage with tarring and chipping was not a good strategy in the longer term, according to Frank Fahey TD, as it would cost more money than doing a proper tar macadam job now (Minahan, 2010).

BUS, TRAIN AND AIR TRAVEL DISRUPTION

Bus services were affected considerably; both city, commuter and long distance services. Train services were barely affected by the cold spell, although there were some problems with signalling at times due to the cold. On 22 December, bus services were affected in Cos. Limerick and Kerry. On 1 January, Bus Éireann suspended services to Sligo and Ballina because of hazardous road conditions around Longford. Dublin Bus had a tough time of it, despite Dublin's roads being gritted daily, and had to cancel all services on the morning of 1 January. By 5pm there was still only a limited service operating, while services to outside Dublin city only operated after midday. Dublin Bus had to cancel more of its services on 2 January. Again on 6 January, all Dublin Bus services were cancelled, and on 12 January it had to reduce commuter services to north Wicklow, Dundrum and Sandyford as a result of the snow.

On 22 December, there was disruption to flights in and out of Ireland, mostly as a result of the bad weather conditions elsewhere in Europe. However, on 1 January, heavy falls of snow closed Dublin Airport for some of the morning. The airport suffered four separate snow showers on the night of 1–2 January, the last one being particularly severe, causing the airport to be closed for four hours, with fourteen flights cancelled and some trans-Atlantic flights diverted to Shannon. This occurred despite the fact that the meteorological station at Dublin Airport only recorded 1cm of snow. Again on 6 January, following a severe snowstorm in the Dublin area, Dublin Airport was closed for almost five hours, causing delays and cancellations to almost one hundred flights. In retrospect, this was a foretaste of what was to come in April (see below, ch. 11).

THE GRITTING CRISIS

The damage to roads was enormous, and the constant freeze-thaw action of the cold spell generated numerous potholes and other damage to road surfaces. The potholes in some places were more than capable of causing serious damage to cars. Local authorities are liable for the damage and numerous claims will be made against them, which will affect their budget for much needed road repairs.

There were substantial complaints across the country about the lack of gritting of key roads, to which local authorities responded that they were at their limit, gritting as much as they could. This led to local authorities using up their usual winter reserves over a very short space of time. It also meant that they had to be selective about which roads to grit, with most secondary and other minor roads not being treated as a result. Sometimes, gritting of stretches of key roads occurred twice a day or even four times a day. There was also concern about the limited storage space available for grit. Local authorities do not store sufficient grit to deal with this type of prolonged and exceptional cold weather event. The cost-benefit of maintaining a larger supply is an issue that now must be addressed by the local authorities.

The minister for transport, Noel Dempsey, stated that 14,000km of road were gritted every day and sometimes even more than once a day, where needed. This was carried out by 261 gritting lorries and 180 snow ploughs. However, it clearly would be impossible to grit the entire road network of 96,000km. More than 60,000 tonnes of salt was spread on roads, a figure in excess of the total annual average for recent years. Local authorities normally had a ten-day supply at hand, which in the past was more than adequate to meet demands.

It took some time for imports of grit to arrive to ease the situation. The port of Cork received 4,000 tonnes on 4 January, Limerick got 4,000 tonnes on the 6th and Carrickfergus unloaded 3,000 tonnes. This was then distributed to local authorities. An additional 14,000 tonnes of grit was made available on 7 January, but this was a rare resupply, and demand was still exceptionally high. Large supplies of salt arrived on Sunday 17 January, but most local authorities had faced the previous week with critically low supplies. Grit was being found from all sorts of unusual sources, even 200 tonnes of salt destined for a Cork company to cure ham was spread on Dublin streets.

Galway City Council estimated that it spent almost €600,000 in total, or €15,000 per day (sixty tons of salt and grit) on gritting alone in response to the cold spell and they appealed for government money to help with this additional and unprecedented cost. The gritting of roads had started as early as 28 November, and it was proving impossible to grit the more than 600km of footpaths in the city. The council also provided eighty housing estates with grit so that they could do the work themselves to speed up the process (Farragher, 2010).

SCHOOL DISRUPTION

On 22 December, many schools in the midlands were closed until the New Year. The normal re-opening after the Christmas holidays did not take place as usual on 6 January because of the ongoing hazardous conditions. A ministerial order was issued to keep schools closed until at least 13 January, much to the joy of many pupils. As a result of the unexpected early thaw on 11 January, however, the education minister, Batt O'Keeffe, rescinded his order to keep schools closed from Monday 11 to Wednesday 13 January.

THE THAW AND BACK TO FLOODING

Even as early as 30 and 31 December, flood problems started to re-emerge. A major clean-up was needed in Skibbereen due to flash flooding on 30 December, when nearly a metre of water accumulated on Townsend Street and Market Street. The flooding occurred as a result of high rainfall in the area, with 55mm falling in a twenty-four-hour period. It is estimated that €4 million worth of damage was caused. Skibbereen had also been flooded on 19 November, and many people and businesses had only recently recovered from this event before being hit again. There was also extreme flooding in the Leap and Rosscarbery areas. Many roads in this area were badly damaged and the public were asked to take extreme caution when driving in these areas due to both the flooding and the consequent road damage.

When the main cold spell started to break from 11 January onwards, conditions changed, with rising temperatures associated with an Atlantic airflow. This also brought heavy rain which, when combined with melting snow, caused flooding in southern areas. In Co. Kerry, the rainfall on 12 January was exceptional, as a record 58.5mm fell on that day, more than twice the monthly total for Shannon Airport.

A significant flood risk emerged for the southeast part of Ireland in the post-thaw period. Initially, spot-flooding was reported on the N11 northbound at Gorey, Co. Wexford, on 15 January. Roads in Co. Carlow were badly affected by flooding caused by heavy rain. In Tullow, the River Slaney burst its banks on 16 January, causing some flood damage with water heights reaching 1.05m. Flooding also affected fifteen businesses in Enniscorthy, Co. Wexford. This was the second time flooding had affected Tullow and Enniscorthy since the previous November. Centuar Street in Carlow town was also flooded for the second time since the big November floods. Damage occurred to two bridges in Co. Carlow, in and near Clonegal. The floods were caused by rainfall and melting snows on the Blackstairs and Wicklow Mountains. Whole sections of roadways in upland areas were washed away as a result (O'Brien, 2010). Homes were evacuated and roads and bridges were closed in Co. Wicklow after more than 30mm of rain fell overnight, causing severe flooding on 16 January, aided by melting snow and ice. The areas badly affected included Arklow, Ashford, Rathnew and parts of Bray.

WATER SHORTAGES AND BURST PIPES

One of the surprising issues of this cold spell was the emergence of water shortages in many parts of the country. This occurred as a result of a number of factors. Firstly, there were numerous burst pipes, ranging from major mains to household pipes all across the country. Most of these breaks were only discovered when the thaw set in. This meant that there was significant loss of water out of the system on top of what is normally lost due to breakages and leaks. Secondly, many people left taps running in order to prevent freezing, thus wasting huge amounts of water. Thirdly, in response to the fact that many local authorities, including the city and county of Dublin, had to cut off water supplies during the night and reduce pressure during the day in order to preserve stocks as much as possible, many householders took to filling baths and other containers with water, which of course had the net effect of increasing pressure on supply. The very high wastage rates in the existing pipe network throughout the country also contributed.

Some snapshots (below) give an idea of just how widespread the water shortage problem was becoming and how long it was going to last. Water cuts were widespread across the country in order to ration supplies in view of the pressures outlined above and the general lack of rainfall during the cold spell.

On 12 January, there were serious issues with water shortage all across Dublin, and in Cork, Sligo, Leitrim and North Tipperary. Dublin City Council said that demand for water the previous weekend was up by 25 per cent and had exceeded 624 million litres per day. Dun Laoghaire-Rathdown County Council was in the same boat, with demand having risen by 26 per cent. In South Dublin, demand exceeded supply by as much as 15 per cent. Severe water conservation measures were introduced in order to stem the massive over-consumption, including systematic and regular water-cuts, while householders and other water-users were asked to limit their consumption to the minimum.

On 14 January, supplies of water in Cork city were still under pressure and some night-time cuts were ongoing. Finglas in Dublin had no water for its fifth day. There were water-cuts in Kilkenny city because of a burst water main, and ongoing problems with water supply at Goresbridge, Gowran, Kilmurry, New Ross, Skeoughvosteen, Slieverue and Thomastown. Clare County Council stated that demand was up by 20 per cent and it was nearing its capacity in terms of water supply. Shannon had supplies cut off for several hours during the day and overnight and standpipes were being used in Ennis, and in north and west Clare. In addition, 200 householders in Co. Wicklow were still without electricity.

By 19 January, water supply across Ireland was improving, but cuts at night were still necessary in many areas. Water supply in Dublin city was still reliant on tankers in some areas as, despite the fact that water pressure was up slightly, rationing had to continue in order to facilitate the detection and repair of leaks. Up to 80 per cent of burst pipes in Dublin were underground and to repair them all would take at least six months.

Many thousands of homes and other properties throughout Ireland suffered from water damage as a result of the thawing of burst pipes. This led to massive insurance claims, as outlined below, and indicates the scale of the damage caused.

The effects of the cold spell did not really impact on many garden species, most of which are resistant to such events, although cold spells in the New Year have a bigger impact as plants have less sugars in storage as the winter progresses and, as a result, are less resistant to cold. Many plants from semi-tropical and tropical environments were killed by the frosts as, although they may be able to survive one or two nights below their normal minimum temperature range, they were not able to cope with such a prolonged spell (Daly, 2010).

INSURANCE CLAIMS AND THE ECONOMIC IMPACT

The Irish Insurance Federation stated that the cost of claims as a result of the cold spell reached almost €300 million, mostly as a result of damage caused by burst pipes. The economic impact was considerable, and there was hardly a business in the country that was not affected by the cold spell and its aftermath, from the smallest retailer to the biggest companies. It is estimated that some hundreds of businesses closed down as a result of the bad weather, as many were already struggling because of the recession. The prolonged cold spell, especially after Christmas and in early January, when big sales normally take place in most retail shops, affected turnover significantly, as people were unwilling or unable to travel to partake in normal sales shopping. In addition, many restaurants and small suppliers of goods and services were also affected by the cold spell. Many people only shopped locally as a result of the poor road conditions and fears about travelling. For example, on 17 January, cancellations at restaurants were high in Dublin because of dangerous roads and pathways (O'Brien and Battles, 2010).

The Irish Business and Employers Confederation (IBEC) estimates that the cold weather cost around €700 million in lost economic output, of which €200–300 million represents a permanent loss, whereas the rest will be made up. Of this €700 million, around €500 million was lost in the service sector and €200 million in manufacturing. This was partly caused by higher than usual absenteeism, at nearly 13 per cent during the cold spell. IBEC also notes that daily work hours lost in January amounted to 17.4 per cent, sales were down by 18.8 per cent and production was down by 12 per cent.

The bad weather of January caused the fastest reduction in manufacturing since August 2009, with the NCB Purchasing Managers' Index dropping from 48.8 in December 2009 to 48.1 in January 2010. This was a result of the havoc caused by the bad weather on both demand and supply sides, which had a negative impact on output, new orders and supplier performance.

Farmers who lost more than 30 per cent of their crops of potatoes and vegetables as a result of the cold spell are eligible for government compensation in accordance with EU State Aid rules. Losses are assessed on a case-by-case basis and overall impact on each grower is an important factor in determining the level of compensation.

INCREASED ENERGY CONSUMPTION

Gas bills went up by an average of 15 per cent, as demand reached record levels because of the cold spell. Demand was as much as 17 per cent higher for the first week in January in comparison with the same time in 2009. Thursday 7 January broke many utility records, including a new high in gas demand, at 9.37 million standard cubic metres, over 30 per cent higher than the previous record of 7.15 million. This was the fourth time in a week that the record had been broken. Bord Gáis donated €1m to the Society of St Vincent de Paul, to help the poor and elderly to pay their gas bills.

Electricity use also broke records, with households using 4,950 megawatts on 7 January, compared with the previous record of 4,906 megawatts in December 2007. The impact across Europe was phenomenal, putting enormous pressure on gas and electricity supplies as householders tried to keep warm.

THE GOVERNMENT RESPONSE

Throughout the cold spell, there were repeated calls from the opposition for the government to intervene in the developing crisis. This was resisted, as much of the work that was needed was being carried out at local authority level, as these authorities have statutory obligations in activities such as road gritting. This changed on 7 January, when the government finally declared a national emergency. The reason for the change was the duration of the cold spell (which by then had reached its twentieth day), the state of the roads across the country (especially concerns about compacted ice on the untreated – and therefore the majority of – roads in the country) and the national shortage of salt and grit, as well as the state of the pavements and the mounting injuries due to falls and car accidents.

The Emergency Response Committee (ERC), chaired by John Gormley, minister for the environment, was convened and it reported to the government on a daily basis from 8 January. The ERC was briefed to look particularly at issues such as transport, schools, access, supplies and health. The matter of getting additional salt and grit was a key part of their agenda, as reserves were in immediate danger of running out. As a result of many countries being in the same situation as Ireland, international demand for salt and grit was exceptionally high and supplies were hard to find in the short term (McGee and Carroll, 2010).

There was both political and public anger that the minister for transport, Noel Dempsey, was on a family holiday in Malta during the worst of the crisis and was stranded and unable to return. Dempsey subsequently claimed that he was in regular contact with his department about the ongoing cold spell and its impact on transport and, like many other families around Ireland, had used the Christmas holidays to get away to the sun.

Humorously, John Gormley was described as the 'minister for snow'. At times, it looked like the government could do no right. For example, a private company was used to clear the snow from the plinth inside Leinster House on 9 January, the day before the biggest fall of snow, which occurred early on 10 January (Corcoran, 2010).

The Houses of the Oireachtas report (2010) on the weather disasters found that the state was underprepared for disasters of this type, that it was unable to cope and that its response mechanism was complex and without one person or body perceived to be taking overall charge. Despite this assessment, Minister Gormley claimed that Ireland's response to the weather crisis had been better than other EU countries, in that we had a bigger stockpile of salt (ten days worth, as opposed to six in Britain), we managed to keep the primary roads ice-free and no homeless people died. It must be pointed out that local authorities and not central government were responsible for the stockpiling of grit and keeping the main roads usable.

LIFE AS NORMAL

It was impossible for many people to live a normal life, as transport was exceedingly difficult, whether going to the local shops, commuting to work or making longer journeys. Numerous sporting and social events were cancelled throughout the country, some to be rescheduled for later in the year, others not to take place at all. The horse-racing industry was particularly affected, as frozen tracks caused the cancellation of numerous race meetings over the Christmas and New Year period and well into January.

All was not lost though, and the usual Christmas swims took place in Dublin and Galway and, despite the conditions, the annual New Year charity swim at Culdaff, Co. Donegal, also went ahead. Some people in Dublin and elsewhere took to their skis to get around, as there was no need to fly off on a skiing holiday when it could be done outside your front door.

CONCLUSIONS

It was a challenging time for all throughout Ireland during the protracted cold spell and its aftermath, when serious problems of flooding, burst pipes and water shortages occurred. This was especially the case for those who had already been through the flooding of November in Cork, Co. Galway and along the Shannon in particular. It seemed at times as if there was no end in sight to the challenges to be faced, especially when the next wave of extraordinary events began.

CHAPTER 11

Strange events

INTRODUCTION

The flooding and the cold spell were exceptional, each in its own right, and even more so as they occurred within the same six-month period. But nature was not finished with us yet. As unexpected and shocking as the flooding and the cold spell were, even stranger phenomena affected Ireland in the first half of 2010. A rare meteorite strike on Ireland was followed by a volcanic eruption in Iceland which, because of the plume of volcanic ash and the continuing dominance of northerly winds across the country, caused travel chaos. Just when we thought there could be nothing else, there was a small but significant earthquake in Co. Clare.

THE METEORITE

On Wednesday 3 February, Ireland had a visitor from space; not a spaceship or a UFO, but a meteorite described by many as a fireball in the sky. It was visible briefly at around 6pm and must have come charging to the ground in Ireland, assuming that any piece of it survived its fiery journey through the atmosphere. It would have been travelling at immense speeds before colliding with the Earth's atmosphere, ten times faster than a bullet (Carroll, 2010). It is impossible to know what its pre-impact size was, but it could have been as large as a house or as small as a football. The last time that Ireland was in receipt of a meteorite was in 1999, when one was seen in Co. Carlow, so these celestial events are very rare. One fell in Co. Derry in 1969 and there is another reference to meteorites hitting Derry in 1844: 'A number of aerolites fell in the neighbourhood of Killeter, Co. Derry' (GBCO, 1852, p. 339). These are some of the very few records of meteorite falls in Ireland.

Many people mistook the 2010 fireball for a crashing helicopter or aeroplane and there were reports of sightings from all over the country, including the midland counties and Cos. Kerry, Cork, Limerick, Armagh and Tyrone. The best view of it was probably obtained by Joss Scott, who was driving in the Glenshane Pass in Co. Derry when he spotted it and described it as being very bright green with an orange tail. Others described it as having a white light. My own view of it was while walking to my car after work on the NUI Galway campus. My attention was caught by what I initially thought was an extra-large firework falling from the sky. I noted that it had a green/orange colour at the time. This matches the description by Joss Scott. Only afterwards did it dawn on me that it was something else and when I heard the reports of the meteorite sighting I knew that that was what I had seen.

The big question of course is whether it survived the atmosphere and actually crashed onto the ground. If so, then whoever found it or pieces of it would be very lucky and quite well-off. There is a huge market internationally for meteorites and even a relatively small piece would be worth thousands of euro. The area where it is most likely to have fallen is near Crimlin in Co. Cavan, but possibly also in Co. Donegal (Newenham, 2010). So far, no one has found any remains of the meteorite and no obvious impact crater has been discovered. If it did reach the ground and crashed into a river or a lake or a bog, the chances of it being found are virtually nil. There is also some suggestion that it may have fallen either in Co. Armagh or in Lough Neagh, as the meteorite was travelling from south to north over the country. Again, there would be no chance of finding it if it fell in Lough Neagh.

THE VOLCANIC ASH CRISIS

Iceland is a remarkable country. It sits on top of the Mid-Atlantic Ridge, which is separating and moving apart as fast as your finger nails are growing. As a result, the island of Iceland is composed of layer upon layer of lava and volcanic ash accumulated over the last twenty million years or so. It is unsurprisingly very volcanically active, with an eruption on average every 3–5 years or so, but there can be bigger gaps and at other times more concentrated activity. Since AD900, Iceland has had 205 eruptions and since 1900 it has had forty-one, the most active volcanoes being Krafla, with nine eruptions since 1900, Askya and Grímsvötn with eight each, followed by Hekla with five.

Icelandic volcanic eruptions almost never impinge on Ireland or northwest Europe, however, and most go unnoticed and barely make it into the press. This is down to the nature of the eruption. Few produce large, long-lived ash plumes, and normal weather patterns tend to keep the ash away from northwest Europe. This is not always the case, however, and as the list of historical cold periods shows, volcanic activity can cause a dramatic cooling in European temperatures, associated with volcanic aerosols reducing incoming solar radiation for anything up to three years after the eruption. Rarer again are ash plumes and falls on Ireland and northwest Europe, although they *have* occurred, albeit not since the development of commercial air traffic. So, the two have never been combined before in a European context, where upwards of 28,000 flights take place each day. It was only a question of time before the two met.

It is very likely that the last time an Icelandic volcano had any significant effect on Ireland was in 1783 during the Laki fissure eruption, which caused devastation all across Iceland and poisonous fogs and aerosols affected Europe at ground level. It is estimated that two million people died, either directly due to the gases or indirectly due to famine associated with crop failures. No study has ever been done to investigate the impact it had on Ireland. The Tambora volcano in Indonesia erupted in 1815, causing such a global downturn in climate that 1816 was known as 'the year with no summer'. Major food crises across the world caused many millions of people to die. The volcano also inspired Mary Shelley's *Frankenstein*, such was the darkness and strangeness of the weather, particularly noticeable in the summer time, as she and her friends were forced to stay indoors all summer in Switzerland. To pass the time, they had a competition to invent a ghost story and *Frankenstein* was what Shelley came up with (Sunstein, 1991). The Krakatoa eruption in 1883 also inspired another major piece of art in the form of Edvard Munch's *The Scream*, painted in 1891 but inspired by the remarkable sunsets caused by this eruption, which were blood red and other bright, unusual and disturbing colours (Olson et al., 2004).

As a result of our own weather crises, nobody paid much attention to a small eruption at Eyjafjallajökull in Iceland in March 2010. The eruption was short-lived and although locally interesting, it did little damage and did not have much impact. It barely made some of the European newspapers. The eruption petered out and few realised that something very dramatic was about to take place. A quick look at the reports of the Icelandic Meteorological Office, which has responsibility for seismic and volcano monitoring, showed that they had been monitoring this area since 1991 because of swarms of shallow earthquakes. These are often indicative of magma moving up to the surface and swelling the earth on top of it. There was a very active swarm of earthquakes in late 2009, which heightened the likelihood of an eruption. The challenge of predicting the actual eruptions is that these earthquake swarms can occur for decades before the eruption finally takes place. The earthquakes also indicated that this small eruption was not going to be the end of the activity in this area (Hjaltadóttir et al., 2009).

In reality, the very minor eruption was little more than the volcano clearing its throat, because on 14 April a large new eruption under the ice-cap of Eyjafjallajökull pumped vast amounts of steam and, more importantly, volcanic ash into the atmosphere (pls 15, 16). This ash consisted primarily of very tiny shards of glass, but also sand and small particles of rock. The weather was such that, as a result of the ongoing cold spell, the prevailing winds were from a northerly and north-westerly direction and these winds pushed the rapidly developing ash plume straight towards Ireland.

As planes cannot fly through volcanic ash plumes for fear of catastrophic engine failure, electrical damage and the sandblasting of the cockpit, there was a massive shut-down of air transport over most of Western Europe. Most people accepted that the shut-down was on health and safety grounds, but it did not help with the travel problems of hundreds of thousands of people. It is fair to say there was massive chaos across Europe, with a knock-on effect throughout the world. Nothing like this had been seen since 9/11 in the USA and not before that since the Second World War. Although the initial no-flight disruption lasted just six days, the backlog of people trying to get home lasted for another two weeks. It was not until the wind direction changed and blew the plume away from Europe that the initial no-fly policy came to an end in some areas, including Ireland.

After the initial shut-down, the rules governing flying in these conditions were changed in response to a better understanding of the way modern jet engines operate in volcanic plumes and improved monitoring of the volcanic plume as it was blown about by the weather patterns. There was ongoing disruption at times, including 4 May, when the wind switched back to the north or northwest and blew the plume back over Ireland and Europe. Fortunately, the eruption ceased on 24 May, much more quickly than the last time it erupted. In 1821, it carried on erupting for nearly fifteen months.

The initial eruption of 2010 was a disaster for individuals and groups trying to travel or get home, as thousands of Irish people were stranded abroad – some for more than a week after their initial departure date. Some missed important family, business or other activities. For the airlines, it was much worse, as under EU legislation they are required to provide compensation to customers who have their flights postponed or cancelled, and this includes paying for accommodation and other expenses. The amounts involved escalated at an exponential rate. The estimate for the initial shutdown is that it cost airlines around €1.1 billion, and each additional shutdown adds to that bill. Airlines and travellers became more efficient at dealing with the shut-downs as they occurred. Irish airlines have lost around €24 million in relation to flights to and from Ireland. Ryanair claims losses approaching €50 million across Europe. In addition, it is estimated that the shut-down across Europe cost passengers an additional €12 million. For many people, the biggest impact was missing important family events, including weddings, and this caused a lot of hurt.

Of course, there are some benefits, particularly for surface transport companies and particularly the ferries, which significantly increased activity and revenue as a result of the initial shut-down and amid people's fears about not being able to fly for important events and holidays. The Government Taskforce on Emergency Planning met in Dublin to discuss the ongoing aviation crisis and its impact on Ireland, but there was little they could do apart from trying to manage the impact, especially when there was ongoing travel disruption.

The last three times the Eyjafjallajökull volcano erupted it triggered off an eruption at the much bigger Katla volcano, some 21km away. There is no sign as yet of Katla erupting however. The two maybe connected by an underground magma conduit, or the ground stress near the Eyjafjallajökull volcano as it erupts may help trigger Katla. Worryingly, Katla would also generate a massive ash cloud, because it is buried under nearly 650m of ice in Mýrdalsjökull and this would make the eruption worse. The ice adds to the explosiveness of the eruption, just in the same way as if you add water to a chip-pan fire. Instead of putting the fire out, it will cause a large flash of flame.

There is some suggestion that worse may be yet to come. This is based on the assessment of the 205 eruptions which have occurred in Iceland since around AD900, when the Vikings began colonising the island (although there were Irish monks on islands off the southeast coast in isolated monasteries before this). The evidence suggests that there are periods when volcanic activity is much greater, and times when activity is relatively quiet. The suggestion is that we are coming to the end of a roughly fifty-year quiet time and moving into a more active phase, in which eruptions will be more frequent and will have an impact on Ireland and Europe. During the last active phase, there was little commercial air traffic, so it will be very different this time. This increased activity may be due to movements in the earth's crust underneath Iceland, creating massive subterranean stresses and it is these stresses that will lead to increased volcanic activity. However, this idea is viewed as controversial and not all vulcanologists agree with it (Leake and Hastings, 2010). Only time will tell.

THE CO. CLARE EARTHQUAKE

Something very strange and probably unique within living memory occurred in Co. Clare on 6 May at 10.24pm. It was a small earthquake or tremor which reached 2.7 on the Richter Scale. This was small by world standards but it was the most significant earth tremor to have affected Ireland since 1984, when Dublin had a magnitude 5.5 earthquake, the biggest ever recorded in Ireland. More importantly, it was the first earthquake to have occurred in Co. Clare since seismic monitoring began in Valentia Observatory in 1978 (Figure 11.1). It was also ten times stronger than the 1.6 magnitude earthquake that struck the Inishowen Peninsula in Co. Donegal on 6 January (Lambkin, 2010). The epicentre of the earthquake was near Lisdoonvarna, but it was felt in many parts of the county, including Liscannor, Lahinch and Doolin.

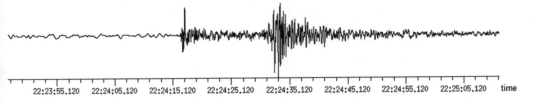

Figure 11.1 Seismograph of the Clare earthquake, Valentia Observatory, Thursday 6 May 2010 (image courtesy of Met Éireann).

The actual shaking associated with the earthquake was over in between three and five seconds, and this is a reflection of the size of the tremor. One of the other main aspects of the earthquake was that it was big enough that people could actually experience it. Noticeable earthquakes only occur in Ireland along the southeast coast up as far as Dublin and in Co. Donegal where there are a number of large and active fault lines. The tremor has woken scientists up and they now say there is a need to re-evaluate the geology of the west of Ireland. Given the short length of records, back only to 1978, it is likely that tremors like this one are more common in the west of Ireland than was previously thought. It also must be stated that this earthquake has no links with what happened in Iceland.

Many people noticed a very loud bang associated with the earthquake and some were concerned enough to phone Garda stations, for fear that it was some kind of explosion. Michael Vaughan, a Lahinch hotel-owner commenting on the noise, stated that 'the sound was something like a sonic boom. Other people living in the area were disturbed by it' (Deegan and Duncan, 2010, p. 3). Many people noticed the shaking, the rattling of crockery, glasses and objects on tables and the swinging of electric lights in rooms and some structural movement. Martin Doyle, who resides in Liscannor, stated that he 'felt the whole roof shake. I thought the wall was collapsing. My neighbour called me and he thought that the sound he heard was me falling down the stairs' (Deegan and Duncan 2010, p. 3). It is likely that the shaking was much more noticeable because the earthquake occurred at the surface and not at some depth beneath the ground. The tremor certainly gave many people a serious fright. It remains a topic of conversation in Co. Clare and no doubt will go down in the history books and in folklore.

CONCLUSIONS

The incredibly rare combination of these three events reminds us that the Earth and the Solar System in which we live are incredibly active, both internally and externally, and we are still only starting to understand how our own planet operates. It reminds us in Ireland that we are not as immune from strange geophysical events like volcanoes and earthquakes as we would like to think. In our lifetime, we may never see this remarkable coincidence again.

CHAPTER 12

Final comments

INTRODUCTION

From summer 2009 to summer 2010, Ireland experienced a sequence of weather and geophysical events unheard of in the long history of the country. A third wet summer in a row was followed in November by extensive flooding across the country, then a major cold spell while parts of the country were drying out, then a probable meteorite strike, an Icelandic volcanic eruption which had a huge impact on air travel, and finally an earthquake in Co. Clare in an area where this was thought to be extremely improbable.

Interestingly, this sequence of events tells us a lot about the Ireland of today in a post-Celtic Tiger era. In many respects, we have been repeatedly tested by the events that occurred. There are a number of results which have emerged from this prolonged test and this final chapter will outline the results in a critical way and the challenges that face us.

In terms of overall impact, these events have cost Ireland over a €1 billion in damages, although the true figure will never be known, as only around half of this was insurable damage. Given that extreme events, particularly flooding, are likely to be more common as the century progresses, this may not be the only time this figure is reached or exceeded. If a similar level of flooding had affected the eastern half of Ireland, then this figure would be easily exceeded given the higher population density and the scale of housing development, especially those developments on flood plains.

THE FLOODING

The wet summers fall outside the current predictions for climate change for Ireland in that the expectation is that we would have reduced summer rainfall, and it is unclear whether this is a trend that will continue or one that represents a blip before drier summers establish themselves.

The flooding of November was probably the biggest flood event to have affected Ireland in over a hundred years, particularly in the south and west.

The record November rainfall is in keeping with the predictions for climate change for Ireland and floods of this type will occur more frequently as the century progresses, not on an annual basis and maybe not even on a decadal basis, but more often than previously.

It has drawn attention to a number of very significant outcomes of the Celtic Tiger building boom. There has been huge development, consisting mostly of housing estates but other buildings and infrastructure as well. Much of this building took place on flood plains. This has been cruelly brought to the attention of many thousands of people who saw their homes flooded, whether in rural or urban areas. This begs the questions, how was this unsuitable land zoned for development, how did developers see this land as suitable for building, how did they get planning permission and what sort of flood assessments were carried out on these sites? Clearly, there was systematic failure, whether deliberate or accidental, in all of these areas, otherwise thousands of new and nearly new homes and other buildings would not have been flooded.

The question now is what to do with these homes, which are vulnerable to future flood events and for which the residents are unlikely to get flood insurance into the future. Who will pay them compensation, given that many of the developers involved are effectively bankrupt. One solution would be to use the National Asset Management Agency (NAMA), which has been put in place to deal with the collapse of the property market and its impact on developers and the banks who lent them the money. It might be possible to move residents from areas at risk of flooding to some of the many ghost estates around the country, as close as possible to their original homes. Then the most vulnerable houses could be demolished and this in turn would help to reduce the oversupply of housing and also maximise as much as possible the housing stock that has been built and is unlikely to be sold in the near future. It would also encourage the completion of some ghost estates. There would be benefits all round.

The demand for flood relief schemes has escalated dramatically as a result of the November floods and the subsequent floods in many parts of Ireland in January. Flood relief schemes for urban areas are expensive: for example, in the case of Enniscorthy, the cost is estimated at around €30 million. The government has upped the available money for flood relief from €43 million to €50 million per year, but this is not nearly enough to meet demand and is only capable of funding a small number of schemes, which typically run over two to three years. This is despite the fact that most flood relief schemes take several years to complete and therefore the funding in each year can be spread out over a number of schemes

(O'Brien, 2010). The flooding has also raised the issue of how we manage our rivers and other waterways in the context of silt removal, something which has not occurred in Irish rivers for over a generation and which may have contributed to the flooding in some locations. The role of the ESB and its operation of hydro-electric stations at Inniscarra and Parteen has also been the cause of much debate and controversy.

The real heroes of the flooding were the ordinary people of Ireland who responded so brilliantly during the flooding, the many voluntary agencies, the army, who contributed enormously, and the much-maligned public sector staff of the local authorities, where the biggest burden fell. National government's response was to provide a limited amount of money and appeal to the EU for help. The national Emergency Response Committee met, but most people saw this as of little value in the real work needed on the ground.

The Irish Insurance Federation did not pull any punches when poor infrastructure, planning and management on the ground were partly to blame for the severity of the problems. There is a clear need for a flood liaison and advice group to manage the situation, considering the likelihood of severe flood events becoming more common (anon, 2010).

THE COLD SPELL

The cold spell, caused primarily by the unusual behaviour of the Arctic Oscillation, descended on us even before the flooding crisis had ended and it again exposed Ireland's lack of preparation. In particular, the low levels of grit stored throughout the country for maintaining road travel caused huge problems off the major roads. Admittedly, the severity and the duration of the cold spell had not been experienced in Ireland since the 1960s and before that the 1940s, so complacency had set in.

By no means does this cold spell negate concerns about global warming, just as one exceptional hot spell does not prove it. As Gibbons (2010) points out, the World Meteorological Organization (WMO) stated that the decade from 2000 to 2009 was the warmest decade on record since widespread instrumental records began *c.*1860. 2009 was the fifth warmest year on record, and each of the thirteen warmest years on record globally occurred since 1996. In fact, global temperatures for the first six months of 2010 were well above the long-term average and, if this trend continues, it will be the hottest year on record. This is partly attributable to El Niño. Gibbon (2010) concluded that people must be made aware of the underlying warming trend, irrespective of this cold weather event and other weather events. In climate change, it is the trend that matters and not the day-to-day weather, despite how dramatic it might be.

But what of global warming? Of course you cannot relate one particular weather event, even one of this nature, directly to global warming, which is concerned with the long-term changes over decades and centuries. There is always going to be year-to-year variability, and since we have had a very long run of very mild years, then a colder winter is always likely once in a while. Even in North Africa, snow might fall once every fifty years or so. However, in an Irish context, the frequency of these cold spells will decline as the century progresses and as the overall warming progresses. This is already occurring, and is witnessed in the general decline in frost days and days with snow over the last forty years or so, and the increase in the length of the grass-growing season by several weeks.

Again, the response to the cold spell fell heavily on the shoulders of the ordinary people of Ireland and the voluntary agencies and the local authorities. One of the most shocking aspects of the cold spell was the thousands of people who suffered broken bones and other serious injuries, primarily because of falls on icy pavements. Much of this could have been avoided if people were obliged by law to clear ice on the pavement in front of their home or business, without liability. This did not occur because you could be sued if you cleared the ice and someone fell on that section. As yet, legislation to change this has not been passed despite its obvious necessity. Huge damage was again caused to houses all over Ireland, primarily by burst pipes once the thaw set in, and this was accompanied by water shortages as people left taps running in order to prevent this, when most did not realise that a dripping tap would be just as effective and would only use a fraction of the water. This crisis also exposed our ongoing problems with water wastage throughout the mains networks, with unaccounted losses exceeding 50 per cent in some areas, mainly due to the age of the pipe network and unrepaired leaks.

There was also great concern among local authorities about trying to get extra funds to cover the cost of repairs to roads and other key infrastructure, given the scale of the damage from both the flooding and the cold spell. In the prevailing economic conditions, no additional monies were forthcoming from central government, who only really became involved when the lack of grit around the country threatened to bring the country to a complete halt. As it was, vast rural areas had to endure ungritted roads for weeks, isolating many individuals and communities.

The failure to understand the concepts of weather and climate change is emphasised by the comments of Noel Ahern TD, who stated that people might have been lulled into a false sense of security by the run of mild winters that have occurred over recent years and said 'I think we might be over-believing this airy-fairy global warming stuff' (Minahan, 2010, p. 1). The public and press disquiet over the perceived lack of government response led the minister for transport, Noel Dempsey, to state that a review was being undertaken at both central and local government level to assess the response to the crisis and whether emergency co-ordination structures should have been activated earlier (Minahan, 2010). The Houses of the Oireachtas report (2010) shows clearly that Ireland is underprepared for major weather disasters and that our management structures are highly complex and confusing and lack core accountability and that more should have been done earlier by central government. The changes that are needed are all the more urgent as severe weather events, particularly floods, storms and higher temperatures and even landslides, are likely to become more common as the century progresses, with reduced intervals between major events (Kiely et al. 2010). Cold spells should gradually become less frequent and less severe as time goes on.

STRANGE EVENTS

If the flooding and cold spell were all that happened over twelve months, this would have been exceptional enough, but nature threw three more events into the mix. These were the probable meteorite strike, the first since 1999, the Co. Clare magnitude 2.7 earthquake (the first recorded there since instrumental records began in 1978) and the volcanic eruption at Eyjafjallajökull. The eruption shut down air travel several times over much of Western Europe, including Ireland. The main shut-down caused massive travel chaos for hundreds of thousands of people throughout the world and some did not get home until two weeks after they were due to, such was the backlog that developed. The cost to the airlines ran into billions. This event reminded us just how isolated we are on the island of Ireland and that without air travel, travel to Europe is totally dependent on ferry transport. Maybe it is time to seriously consider a land bridge from Ireland to Britain. This would provide massive employment and in the long term would be the most important piece of infrastructure built for Ireland. This is technically feasible but the cost would run into billions of euro.

CONCLUSIONS

There is no doubt that the biggest warnings of the events of 2009–10 are that we are unprepared for weather-related crises and that much development that took place during the Celtic Tiger is in the wrong place. The planning system failed to prevent this happening, much to the cost of ordinary people who bought houses in estates in good faith. We also need stronger and better financed local authorities, as these are the ones that deal with the real problems on the ground.

But more than anything else, the events discussed in this book show that Ireland has not lost its sense of community and the real heroes were the ordinary people who responded magnificently to the crises as they unfolded, whether individually or through voluntary services and agencies. This book in some respects is written as a tribute to everybody who helped, whether it was making a cup of tea for someone, providing a shoulder to cry on, filling sand bags, taking people into their homes, making a contribution of cash, goods or services, or any other activity that helped when the need arose. Finally, at the time of going to press, it must be noted that not all people have had their lives restored to what they were before, and some people are still struggling with the impact of these weather crises.

Bibliography

Anon. 1960. '*Innisfallen* in Cork over 7 hours late', *Cork Examiner*, 5/12/1960.

Anon. 2009. 'Roofing blown off Dublin apartments', *Irish Times*, 25/11/2009.

Anon. 2010a. 'Hottest year on record, figures show', *Irish Times*, 17/7/2010.

Anon. 2010b. 'Big freeze and flood claims at €541m', RTE Business, 25/2/2010
(www.rte.ie/business/2010/0225/insurance, accessed 14 July 2010).

Barker, T. and E. Cassidy. 1988. 'The Venice of the north ...', *Cork Examiner*,
26/10/1988.

Bedford, R. 2008. *Yesterday's weather*. Self-published.

Beesley, A. 2010. 'Cold snap death toll rises across Europe', *Irish Times*, 8/1/2010.

Boate, G. 1652. *Irelands naturall history being a true and ample description of its
situation, greatness, shape, and nature, of its hills, woods, heaths, bog, of its
fruitfull parts, and profitable grounds : with the severall ways of manuring and
improving the same : with its heads or promontories, harbours, roads, and bays,
of its springs, and fountains, brooks, rivers, loghs, of its metalls, mineralls, free-
stone, marble, sea-coal, turf, and other things that are taken out of the ground :
and lastly of the nature and temperature of its air and season, and what diseases
it is free from or subject unto : conducing to the advancement of navigation,
husbandry, and other profitable arts and professions.* London.

Booth, G. 2010. 'Severe wintry weather: 1947', *Weather*, 65:3, 58.

Britton, C.E. 1937. *A meteorological chronology to AD1450*, Meteorological Office
Geophysical Memoirs, 70.

Carroll, S. 2010. 'Space rock proves an atmospheric fireball', *Irish Times*,
4/2/2010.

Chabot, J.-B. 1899, 1901, 1905, 1910, 1924. *Chronique de Michel le Syrien,
patriarche Jacobite d'antiche (1166–1199). Éditée pour la première fois et
traduite en français, Tome I–V.* Paris.

Clube, V. 1992. 'The fundamental role of giant comets in Earth history', *Celestial
Mechanics and Dynamical Astronomy*, 54, 179–93.

Corcoran, J. 2010. 'Big freeze shambles as bad as Dad's Army', *Sunday
Independent*, 10/1/2010.

Daly, G. 2010. 'Plants face struggle to survive freezing fallout', *Sunday
Independent*, 10/1/2010.

Deegan, G. and P. Duncan. 2010. 'West Clare hit by seismic tremor', *Irish Times*,
8/5/2010.

Dawson, A. 2009. *So foul and fair a day: a history of Scotland's weather and climate.*
Edinburgh.

Dickson, D. 1997. *Arctic Ireland: the extraordinary story of the great frost and
forgotten famine of 1740–1741.* Belfast.

Eddy, J.A. 1980. 'Climate and the role of the sun', *Journal of Interdisciplinary History*, 10:4, 725–47.

Eden, P. 2008. *Great British weather disasters*. London.

Farragher, F. 2010. 'Government aid sought for 600k freeze spend', *Galway Tribune*, 15/1/2010.

GBCO (Great Britain Census Office). 1852. *Census of Ireland, 1851: supplementary report on tables of deaths*. London.

Gibbons, J. 2010. 'Cold "snap" does not undo climate trends', *Irish Times*, 14/1/2010.

Hayes, K. 2009. 'Concern over record water levels: Gormley', *Irish Times*, 27/11/2009.

Heery, S. 1993. *The Shannon floodlands: a natural history of the Shannon callows*. Kinvara.

Hickey, K.R. 1990. 'The historical climatology of flooding in Cork city from 1841–1988'. MA, Geography, UCC.

Hickey, K.R. 2005. 'Flooding in the city' in J.S. Crowley, R.J.N. Devoy, D. Lenihan and P. O'Flanagan (eds), *Atlas of Cork city*. Cork, 25–31.

Hickey, K.R. 2008. *Five minutes to midnight? Ireland and climate change*. Belfast.

Houses of the Oireachtas. 2010. 'The management of severe weather events in Ireland and related matters'. Joint Committee on the Environment, Heritage and Local Government, Fourth Report, July 2010.

Hjaltadóttir, S., K.S. Vogfjörð and R. Slunga. 2009. 'Seismic signs of magma pathways through the crust in the Eyjafjallajökull volcano, South Iceland'. Reykjavik.

Kelleher, O. 2010. 'Cork flood defences could cost €100m', *Irish Times*, 2/2/2010.

Kiely G., P. Leahy, F. Ludlow, B. Stefanini, E. Reilly, M. Monk and J. Harris. 2010. 'Extreme weather, climate and natural disasters in Ireland'. Environmental Protection Agency, Climate Change Research Programme, 2007–2013, Report Series, 5.

Kinver, M. 2010. 'Sun activity link to cold winters', BBC News, 15/4/2010 (www.news.bbc.co.uk/2/hi/science/nature/8615789, accessed 14 July 2010).

Lambkin, K. 2010. 'Seismic tremor in Co. Clare', www.met.ie/news/display.asp?ID=66 (accessed 10 January 2010).

Leake, J. 2010. 'El Niño could make 2010 the hottest year ever', *Sunday Times*, 23/5/2010.

Leake, J. and C. Hastings. 2010. 'Volcanic ash chaos "could last decades"', *Sunday Times*, 16/5/2010.

Lennon, P. and S. Walsh. 2008. '2008 summer rainfall in Ireland', *Climatological Note*, 11, Dublin.

Lowe, E.J. 1870. *Natural phenomena and chronology of the seasons: being an account of remarkable frosts, droughts, thunderstorms, gales, floods, earthquakes, etc also diseases, cattle plagues, famines etc, which have occurred in the British Isles since AD220, chronologically arranged*. London.

Manley, G. 1974 (and updates). 'Central England temperatures: monthly means, 1659 to 1973', *Quarterly Journal of the Royal Meteorological Society*, 100:425, 389–405.

Martin, G. 2009. 'Impact of flooding', *Irish Times*, Letters page, 24/11/2009.

McCaffrey, U. 2010. 'Aviva blames weaker economy, weather for 25 per cent fall in profits', *Irish Times*, 5/3/2010.

McDonagh, M. 2009. 'Town risks being cut off by flooding', *Irish Times*, 20/11/2009.

McGee, H. and S. Carroll. 2010. 'Government intervenes to co-ordinate response to big freeze', *Irish Times*, 8/1/2010.

McGreevy, R. 2010a. 'Parts of country got November rainfall that would only occur every 500 years', *Irish Times*, 17/2/2010.

McGreevy, R. 2010b. 'State to seek flood relief funds from EU', *Irish Times*, 22/1/2010.

McGreevy, R. 2010c. 'Cold spell set to become the longest in over 40 years', *Irish Times*, 2/1/2010.

McGreevy, R. 2010d. 'Low of -16°C recorded in coldest winter for 47 years', *Irish Times*, 2/3/2010.

McWilliams, B. 2009. *Weather eye: the final year.* Dublin.

Melia, P. and R. Reigel. 2010. 'ESB was warned dams could not cope with flood', *Irish Independent*, 11/5/2010.

Met Éireann. 2006. June, July and August monthly weather bulletins, 242, 243, 244.

Met Éireann. 2007. June, July and August monthly weather bulletins, 254, 255, 256.

Met Éireann. 2008. June, July and August monthly weather bulletins, 266, 267, 268.

Met Éireann. 2009. March, June, July, August, September, October, November, December monthly weather bulletins, 275, 278, 279, 280, 281, 282, 283, 284.

Met Éireann. 2010. January, February, March, April, May, June monthly weather bulletins, 285, 286, 287, 288, 289, 290.

Minahan, M. 2010. 'Householders may be forced to clear snow', *Irish Times*, 21/1/2010.

Newenham, P. 2010. 'Meteorite "may have landed in Donegal"', *Irish Times*, 9/2/2010.

O'Brien, S. and J. Battles. 2010. 'Big freeze chills hope of business recovery', *Sunday Times*, 10/1/2010.

O'Brien, T. 2010. 'Carlow and Wexford hit by renewed flooding', *Irish Times*, 19/1/2010.

Office of Public Works. 2009. *The planning system and flood risk management: guidelines for planning authorities.* Dublin.

Olson, D.W., R.L. Doescher and M.S. Olson. 2004. 'The blood-red sky of *the Scream*', *American Physical Society News*, 13:5 (also www.aps.org/publications/apsnews/200405/backpage.cfm, accessed 11 August 2010).

Deluge

Parsons, M. 2010. 'Councillors' planning decisions "catastrophic"', *Irish Times*, 25/1/2010.

Roche, B. 2010. 'Plan to manage River Lee levels welcomed', *Irish Times*, 3/3/2010.

Rohan, P.K. 1986. *The climate of Ireland.* 2nd ed., Dublin.

Rydell, L.E. 1956. *River Shannon flood problem: final report by L.E. Rydell, Corps of Engineers, US Army.* Dublin.

Siggins, L. 2009a. 'Floods cause traffic chaos in the west', *Irish Times*, 23/11/2009.

Siggins, L. 2009b. 'Hundreds of homes still under water in the west', *Irish Times*, 3/12/2009.

Siggins, L. 2010. '"High-vis" jacket key to rescue of elderly farmer', *Irish Times*, 2/1/2010.

Simkin, S.L. 2002. *Volcanoes of the world: an illustrated catalogue of Holocene volcanoes and their eruptions.* Washington DC.

Smith, C. 1839. *The ancient and present state of the county and city of Cork, containing a natural, civil, ecclesiastical, historical and topographical description thereof.* 2 vols, Cork.

Sunstein, E.W. 1991. *Mary Shelley: romance and reality.* Baltimore MD.

Tempest, H.G. 1944. 'The monastery of Inis-Mocht', *Journal of the County Louth Archaeological and Historical Society*, 10:4, 342–5.

Tuckey, F.H. 1837. *The county and city of Cork remembrancer; or, annals of the county and city of Cork.* Cork.

Tyrrell, J.G. and K.R. Hickey. 1992. 'A flood chronology for Cork city and its climatological background', *Irish Geography*, 24:2, 81–90.

Walsh, S. 2010. 'Report on the rainfall of November 2009', *Climatological Note*, 12, Dublin.

Wohletz, K.H. 2000. 'Were the Dark Ages triggered by volcano-related climate changes in the 6th century?', *Transactions of the American Geophysical Union*, 48:81, F1305.

Index